Fuel for Change: World Bank energy policy – rhetoric and reality

edited by Ian Tellam

Zed Books

LONDON • NEW YORK

Both ENDS

AMSTERDAM

Fuel for Change: World Bank energy policy – rhetoric and reality was first published by Zed Books Ltd, 7 Cynthia Street, London N1 9JF, UK, and Room 400, 175 Fifth Avenue, New York, NY 10010, USA, in 2000.

in association with

Both ENDS, Damrak 28–30, 1012 LJ Amsterdam, The Netherlands

Distributed in the USA exclusively by St Martin's Press, Inc., 175 Fifth Avenue, New York, NY 10010, USA

Cover designed by Andrew Corbett
Set in Monotype Ehrhardt and Franklin Gothic by Ewan Smith
Printed and bound in the United Kingdom by Biddles Ltd, Guildford and King's Lynn

A catalogue record for this book is available from the British Library.

Library of Congress Cataloging-in-Publication Data

Fuel for change: World Bank energy policy: rhetoric and reality / edited by Ian Tellam.
        p. cm.
    Includes bibliographical references and index.
    ISBN 1-85649-781-X (cased) – ISBN 1-85649-782-8 (limp)
    1. Energy development–Developing countries–Finance–Case studies. 2. Energy development–Europe, Eastern–Finance–Case studies. 3. Renewable energy sources–Developing countries. 4. Renewable energy sources–Europe, Eastern. 5. World Bank. I. Title: World Bank energy policy. II. Tellam, Ian.

HD9502.D442 F83 2000
333.79–dc21

                                                          00-027614

ISBN 1 85649 781 X cased
ISBN 1 85649 782 8 limp

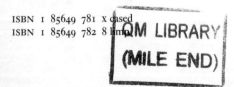

About the Editor

Ian Tellam has worked for ten years on energy, environment and development issues in a number of international non-governmental organizations (NGOs). He is currently a senior researcher in international energy policy with the Centre for Energy Conservation and Environmental Technology in Delft, The Netherlands, and an associate lecturer in international environmental policy with the Open University in the UK.

About Both ENDS

Both ENDS is an Amsterdam-based service and advocacy organization which collaborates with hundreds of environmental and indigenous organizations, both in the South and in the North, with the aim of helping to create and sustain a vigilant and effective environmental movement.

Zed Titles on Energy

So long as climate change and global warming remain one of the most unpredictable environmental threats, energy questions – and, in particular, how to use energy sources more efficiently and how to switch to renewables – will be a hugely important policy arena for both already industrialized and developing countries. Zed Books has published a series of titles addressing various aspects of these questions.

African Energy Policy Research Series: various titles

Bina Agarwal, *Cold Hearths and Barren Slopes: The Woodfuel Crisis in the Third World*

Jean-Claude Debeir et al., *In the Servitude of Power: Energy and Civilization through the Ages*

Stephen Karekezi and Gordon Mackenzie (eds), *Energy Options for Africa: Environmentally Sustainable Alternatives* (in association with the UNEP Collaborating Centre on Energy and Environment)

Patrick McCully, *Silenced Rivers: The Ecology and Politics of Large Dams*

Peter Read, *Responding to Global Warming: The Technology, Economics and Politics of Sustainable Energy*

Wolfgang Sachs, Reinhard Loske and Manfred Linz, *Greening the North: A Post-Industrial Blueprint for Ecology and Equity*

Ian Tellam (ed.), *Fuel for Change: World Bank Energy Policy – Rhetoric and Reality*

Ernst von Weizsacker and Jochen Jesinghaus, *Ecological Tax Reform: A Policy Proposal for Sustainable Development*

For full details about these titles and Zed's general and subject catalogues, please write to: The Marketing Department, Zed Books, 7 Cynthia Street, London NI 9JF, UK or e-mail: sales@zedbooks.demon.co.uk

Visit our website at: http://www.zedbooks.demon.co.uk

Contents

Boxes

Tables

Acknowledgements

A large number of people supported the production of this book in a variety of ways. Ophelia Cowell gave valuable comments and carried out preliminary editing on a number of the country studies. Bert Zijlstra read drafts of the manuscript and provided valuable comments. Ricardo Carrere, Petr Hlobil, Ong'wen Oduor and André Ballesteros were responsible for coordinating the production of the country studies and for providing regional analyses. Luis Stolovich helped to shape the Latin American analysis and Ann Heidenreich copy-edited the first versions of the Latin American country studies. Both ENDS in Amsterdam, especially past and present directors Theo van Koolwijk and Marie José Vervest, gave important support. Past and present staff of Both ENDS helped in many ways, and thanks should go in particular to Rutger Brandenburg, Laurian Zwart, Wiert Wiertsema, Huub Scheele, Daniel von Moltke and Theo Ruyter. Also, Roberto Bissio, Mike Philips, Athena Ronquillo, Anthony Froggatt, Daphne Wysham and Susan George provided valuable comments on early drafts of various chapters, which helped to shape the overall analysis. Pim, Trees and Cintha van Hoof, and Daniel and Jiska Tellam, deserve thanks for being flexible with time and space to allow the work to be completed. Kay Tellam has given continuing and valuable support. And Robert Molteno at Zed Books should be gratefully acknowledged for patiently working with the editor and manuscript.

About the Contributing Organizations

Global Village Cameroon, B.P. 3499 Yaounde, Cameroon

Founded in 1996, Global Village Cameroon (GVC) is an NGO registered in Cameroon. GVC operates at both national and regional levels and works to enhance environmental protection and sustainable development. It works on solutions to national and regional problems and is focusing on environmental degradation, poverty alleviation, energy and economy issues, and sustainable human settlements.

Resource Projects Kenya, P.O. Box 42994, Nairobi, Kenya

Resource Projects Kenya is a Nairobi-based NGO that has carried out research into energy and development issues as they relate to sustainable development in Africa.

ZERO, P.O. Box 5338, Harare, Zimbabwe

Established in 1987, ZERO is a coordinating NGO on environmental issues, which conducts research, promotes environmental policy and disseminates information.

EcoNews Africa, P.O. Box 76406, Nairobi, Kenya

Established in 1992, EcoNews Africa is an NGO initiative that strives to link international processes touching on environment and development to local initiatives and experiences.

Center for Energy and Environmental Policy, University of Delaware, Newark, DE 19716, USA

CEEP is the principal academic and research unit for graduate study in the areas of energy and environmental policy at the University of Delaware, USA. The centre supports interdisciplinary and collaborative research and graduate studies in the fields of energy and environmental policy; technology, environment and society; political economy of energy and environment; and sustainable development. Centre research and teaching are informed by theories and concepts drawn from the fields of political economy and science, technology and society. CEEP provided the Chinese country study for this book, because it was not possible to find a Chinese organization that was able to write freely and openly on Chinese energy, environment and development issues.

Prayas, Amrita Clinic, Athawale Corner, Karve Road, Corner Deccan Gymkhana, Pune – 411 004, India

Prayas means determined efforts in a definite direction. Prayas staff apply professional knowledge and skills to understand the issues afflicting society, especially in the areas of health, energy and livelihoods. Further, Prayas staff strive to translate this understanding into strategic but sensitive responses. Prayas believes that, if equipped with adequate information, sound analyses, and the necessary skills, even disadvantaged sections of society can tackle their problems and shape their own futures. Prayas' activities – research, policy analyses, information dissemination, public interest advocacy, skill development, and the provision of counselling support – are geared to the objective of equipping the disadvantaged and facilitating people's own actions.

Pelangi, Janlan Danau Tondano, A-4 Jakarta 10210, Indonesia

Pelangi Indonesia is a non-profit, non-governmental research organization for sustainable development. Established in 1992 by Indonesian environmentalists, it was born out of the need for independent policy studies on sustainable development issues. It is a resource institution and its activities are geared towards strengthening the efforts to amplify people's participation in development projects and issues of sustainable development. Pelangi Indonesia achieves its objectives through rigorous research activities, combined with active collaborations with its stakeholders. Among others, Pelangi Indonesia conducts case study and pilot project analysis, and original action and participatory research.

Legal Rights and Natural Resources Centre, 3/F Puno Building, #47 Kalayaan Avenue, Diliman, Quezon City, Philippines

The Legal Rights and Natural Resources Centre is a policy and legal research and advocacy institution primarily dealing with land tenure rights, indigenous people, forestry, resource management and energy. The main objective is to bridge the gap between local communities and the official policy-making bodies, and to allow people to articulate their aspirations and to have dialogue. Established in 1987, the centre provides legal training, local and national level policy research, and national campaign support to communities affected by destructive development projects.

Bulgarian Centre for Environmental Information and Education, Jk. ILINDEN, Indje voivoda str., Bl. 9 (G1), entr.3, 2 floor, 1309 Sofia, Bulgaria

The Centre for Environmental Information and Education (CEIE) was established in late 1992. CEIE is working on a variety of issues, including public awareness campaigns, environmental education, coordination of local NGO activities countrywide, lobbying, and monitoring of IFIs. It has special campaigns in energy, transportation and water issues. It has been a member of the CEE Bankwatch Network since 1996. In 1998 CEIE became one of the founders of the Bulgarian Association for Alternative Tourism. Since 1997 CEIE, together with the NGO For the Earth, has published a bi-monthly *BankWatch Bulletin* in Bulgarian.

Energy Club, P.O. Box 411, H-1519 Budapest, Hungary

Energy Club (Energia Klub) is an NGO working in close cooperation with various independent environmental organizations throughout Hungary. The mission of Energia Klub is to empower other NGOs with skills and knowledge on campaigning for sustainable energy. One of the latest projects is an Energy Efficiency Programme, which aims to raise public awareness at various levels of society (households, media, schools, etc.). Twenty NGOs have joined this programme since the beginning of 1995. At the end of 1996 Energia Klub started another programme, FAIRE – Free and Applied Internship for Renewables and Efficiency – which is an internship programme for Central Eastern European energy activists and teaches issues related to sustainable energy and public campaigning.

Lithuanian Green Movement, Central Post P.O. 156, LT–3000 Kaunas, Lithuania

The Lithuanian Green Movement (LGM) was established in 1988 and is one of the most influential Lithuanian environmental NGOs. It played a key role in the process of democratization of Lithuania during the period 1988 to 1990. LGM is an umbrella union of environmental clubs, groups and individuals. The main focus of its activities are the protection of the Baltic Sea, conservation of protected territories and natural landscape, an energy campaign that works for decentralization and democratization of the existing energy system, and air pollution and acid rain. Environmental education is an important part of LGM's work, focusing on sustainable development and the creation of a pluralistic, democratic society.

National Ecological Centre, Uritskogo 15–115, Kiev 252035, Ukraine

The National Ecological Centre of Ukraine (EcoCentre) was registered as an NGO in 1991. Its main goal is to link the efforts of scientists, environmental groups and individuals to improve the state of the environment in Ukraine. EcoCentre has 18 branches throughout Ukraine and several daughter organizations.

CEE Bankwatch, c/o Biocit, Chlumova 17, 130 00 Praha 3 – Zizkov, Czech Republic

The CEE Bankwatch Network is an international NGO with member organizations in eleven countries in Central and Eastern Europe (CEE) and the Commonwealth of Independent States (CIS) region. The basic aim of the network is to monitor the activities of international financial institutions (IFIs) in the region, and to propose constructive alternatives to their policies and projects. The CEE Bankwatch Network was formally set up in 1995 and is focusing mainly on energy and transport, while working to promote public participation and access to information concerning the activities of IFIs in the CEE region. Members of the CEE Bankwatch Network attend the annual meetings of the IFIs and are engaged in an ongoing critical dialogue with IFI staff and executive directors at national, regional and international levels.

Centro de Estudos em Energia e Meio Ambiente (CEEMA) Electrotechnics and Energy Institute, University of São Paulo, Av. Prof. Almeida Prado, 925, 05508–900 São Paulo – SP, Brazil

CEEMA is an energy and environment research centre, based at the University of São Paulo. CEEMA has been involved in national discussions in Brazil on a number of issues concerning energy sector privatization, the environmental impacts of energy use, and the promotion of energy efficiency and energy conservation.

CENSAT 'Agua Viva', Carrera 19, No. 29 – 12 O. 202, Apartado Aéreo No. 16789, Telefax: 57 – 1 – 2456860, Santafé de Bogotá, Colombia

CENSAT collaborates with civic, peasant and youth organizations, women's groups, indigenous people, teachers, unions, cultural groups, academics and municipal government bodies. It is active on the social and environmental impacts of mining, the petroleum industry and the multilateral development banks. CENSAT is carrying out research into the effects of horticulture on health and the environment and has a programme of research and implementation of agricultural alternatives. It further advises trade union education and research programmes, and is involved in environmental education.

Equipo Pueblo, Francisco Field Jurado #51, col. Independencia 03630, Mexico

Equipo Pueblo is a Mexican NGO, founded in 1977. It works closely with social organizations and citizen coalitions in the promotion of democracy, social development, defence of human rights, and economic justice.

Centro De Estudios Uruguayo De Tecnologias Apropiadas, Programa de Energias Renovables, Casilla de Correos 5049, Santiago de Chile 1183, 11200 Montevideo, Uruguay

The Uruguayan Centre for Appropriate Technologies (CEUTA) is an independent, non-profit foundation dedicated to the study and promotion of technologies for sustainable development. CEUTA's working methods involve training, technical consultancy and research. CEUTA carries out projects involving international cooperation as well as projects that involve collaboration with national institutions. It has four working programmes: renewable energy; agro-ecology; medicinal plants; and an environmental education and training programme called 'enfoque 21'.

Instituto del Tercer Mundo, Jackson 1136, 11200 Montevideo, Uruguay

The main aim of Instituto del Tercer Mundo (IteM) (Third World Institute) is to contribute to the strengthening of civil society by promoting informed and democratic decision-making, respect for human rights, freedom of speech, education, and access to information by broad sectors of the population and social organizations.

Abbreviations and Acronyms

ADB	Asian Development Bank
ADME	Wholesale Electricity Market Administration
AfDB	African Development Bank
AIJ	Activities Implemented Jointly Program
ANCAP	state-owned oil company (Uruguay)
ANEEL	National Electricity Agency (Brazil)
ANP	National Oil Agency
APEC	Asia–Pacific Economic Cooperation
ASTAE	Asian Alternative Energy Unit
BADEA	Arab–African Development Bank
BNDES	National Bank for Economic and Social Development (Brazil)
BOO	build-operate-own
BOT	build-operate-transfer
BP	Bank Procedures Document (World Bank)
CCGT	combined cycle gas turbine
CEE	Central and Eastern Europe
CELPE	Companhia Elétrica de Pernambuco (Brazil)
CEMIG	Companhia Energética de Minas Gerais (Brazil)
CEPEL	Compañía Paranaense de Energía (Brazil)
CESMP	Coal India Environmental and Social Mitigation Project
CFE	Comisión Federal de Electricidad (Mexico)
CHP	combined heat and power plant
COMECON	Council for Mutual Economic Assistance
CONAE	National Commission for Energy Conservation (Mexico)
CRE	Energy Regulatory Commission (Mexico)
CREG	Regulating Commission of Energy and Gas (Colombia)
DENR	Department of Environment and Natural Resources (Philippines)
DMC	developing member country
DOE	Department of Energy
DSM	demand-side management
EBRD	European Bank for Reconstruction and Development
ECC	Environmental Compliance Certificate (Philippines)
EE	energy efficiency
EESW	Energy–Environment Sector Work
EGF	Environmental Guarantee Fund (Philippines)
EIA	Environmental Impact Assessment
EIB	European Investment Bank

EIS	Environmental Impact Study (Philippines)
ESAP	Energy Saving Action Programme
ESCO	Energy Service Company
ESMAP	Energy Sector Management Assistance Programme
FIDE	Mexican energy conservation trusteeship
FINESSE	Financing Energy Services for Small-Scale Energy Users
FIS	Social Investment Fund (Mexico)
GASEBA	Uruguayan gas company
GEF	Global Environment Facility
GHG	greenhouse gases
GP	Good Practices Document, World Bank
GWh	gigawatt hour
IAEA	International Atomic Energy Agency
IBRD	International Bank for Reconstruction and Development
IDA	International Development Association
IDB	Inter-American Development Bank
IDBI	Industrial Development Bank of India
IEA	International Energy Agency
IFAD	International Fund for Agricultural and Rural Development
IFC	International Finance Corporation
IFIs	international financial institutions
IIEC	International Institute for Energy Conservation
INEA	Instituto de Ciencias Nucleares y Alternativas (Colombia)
IOU	investor-owned utility
IPP	independent power producer
IREDA	Indian Renewable Energy Development Agency
IRP	integrated resource planning
ISA	Interconexion Electrica Nacional (Brazil)
IsDB	Islamic Development Bank
ITDG	Intermediate Technology Development Group
JBEC	Java–Bali Electricity Company
JBTC	Java–Bali Transmission Company
KEAP	Kenya Energy Auditing Programme
KEMP	Kenya Energy Management Programme
KfW	Kreditanstalt für Wiederaufbau
KPC	Kenya Power Company
KPLC	Kenya Power and Light Company
MDB	multilateral development bank
MIEM	Ministry of Industry, Energy and Mining (Uruguay)
MMT	Multipartite Monitoring Team (Philippines)
MWh	megawatt hour
NAFTA	North American Free Trade Agreement
NEK	National Electricity Company of Bulgaria
NG	natural gas
NGO	non-governmental organization
NIC	newly industrialized country
NOCZIM	National Oil Company of Zimbabwe
NPC	National Power Corporation (Philippines)

NPP	nuclear power plant
NSA	Nuclear Safety Account
NTPC	National Thermal Power Corporation (India)
OD	Operational Directive (World Bank)
OECF	Overseas Economic Cooperation Fund of Japan
OED	World Bank Operations Evaluation Department
OP	Operational Policy Document (World Bank)
PAP	project-affected people
PHARE	European Union grant programme for Central and East European countries
PLN	Indonesian State Power Company
PNP	Philippine National Police
PPA	power purchase agreement
PROALCOOL	Alcohol National Programme (Brazil)
PROCEL	National Programme for Electricity Conservation (Brazil)
PSD	Project Summary Document
PSP	pumped storage plant
PVMTI	Photovoltaic Market Transformation Initiative
QPL	Quezon Power (Philippines), Limited Co.
RE	renewable energy
REEF	Renewable Energy and Energy Efficiency fund
ROW	right-of-way agreements (Philippines)
RPTES	Regional Programme on the Traditional Energy Sector
RTC	Regional Trial Court (Philippines)
SADC	Southern African Development Community
SDC	Solar Development Corporation
SDPC	National Corporation of Petrol Stations, Cameroon
SEB	State Electricity Board (India)
SEDESOL	Secretariat for Social Development (Mexico)
SEMIP	Secretariat of Energy, Mining and Industry (Mexico)
SHCP	Treasury and Public Credit Secretariat (Mexico)
SNH	National Hydrocarbon Corporation (Cameroon)
SONEL	National Electricity Corporation (Cameroon)
TDA	United States Trade and Development Agency
TOE	tons of oil equivalent
TPP	thermal power plant
TWh	terawatt hour (1 TW is equivalent to 1,000 GW)
UCPTE	West European Electricity Network
UFPE	Universidade Federal de Pernambuco (Brazil)
UkrESCO	Ukrainian Energy Saving Company
UNDP	United Nations Development Programme
UNFCCC	UN Framework Convention on Climate Change
UPS	Unified Power System
UTE	National Administration of Power Plants and Electricity Transmission (Uruguay)
VIDCO	Village Development Committee
WADCO	Ward Development Committee
ZESA	Zimbabwe Electricity Supply Authority

Part I

The Burning Issues

Fuelling Change

Introduction

The energy sector is of particular importance in economic development. Yet the sector is facing huge problems. There is a lack of finance to meet the costs of providing enough electric power to cover expected energy demand, particularly in the fast-growing economies in Asia. There is great inequality between the excessive amounts of energy used by most people in industrialized countries, compared with the lack of access to basic energy services, such as power for lighting and fuel for cooking, that confronts almost three billion people in low-income countries around the world. There are urgent environmental problems, ranging from the local air pollution caused by inefficient charcoal-burning stoves, which is damaging the health of people in rural areas of Africa, for example, to the global problem of climate change, which is now increasingly recognized as being real, and which threatens widespread social and economic dislocation. Most governments are now agreed that serious action is needed to provide energy services in ways that are both environmentally sound and equitable. This desire is reflected in the energy policies of the world's foremost inter-national financial institution, the lead agency making policy in the field of energy and development: the World Bank. The Bank's policies state the importance of providing energy in an environmentally sustainable and equitable manner. The problem is that this rhetoric is not matched by the reality of the Bank's investments in the global South and in Central and Eastern Europe. That is what this book is about: an assessment from the field, based on a large sample of actual country experiences, of what the Bank is actually doing.

In recent years, a series of independent reports have criticized the World Bank for its poor social and environmental record in the energy sector. These reports have generally included constructive recommendations for the promotion of sustainable energy by the Bank. However, on the occasions when it has responded to these reports the Bank has tended to downplay their importance on the grounds that, for example:

- the reports contain factual errors;
- the data in the reports are out of date;
- the methodology used by the authors is inadequate;
- the authors do not properly understand the Bank's work in relation to energy efficiency/rural energy/renewable energy;
- the authors do not properly understand internal procedures in the Bank;
- the authors (come from rich countries and) do not properly understand the energy needs of poor people in low-income countries;
- the authors are politically motivated, have an interest in attacking the Bank, and are playing a 'cruel hoax' on an unsuspecting public.[1]

Fuel for Change compares the Bank's policies and rhetoric with facts and experiences as collected and evaluated by non-governmental organizations (NGOs) from Latin America, Africa, Asia and Central and Eastern Europe. The title is a play on words, since the Bank's latest energy sector strategy paper is called *Fuel for Thought.* Judging by past experience, it is most likely that if the Bank responds to this book at all it will attempt to discredit the authors and the findings using one or all of the above arguments.

It is important to explain at this point, therefore, that while every attempt has been made to ensure that all the facts and figures used are correct, this work is not intended as an academic exercise, but rather – as the title indicates – as fuel for change. The book does not test a hypothesis, but is rather a discourse that examines the issues surrounding the World Bank's level of commitment to sustainable energy. Its conclusions are based on studies of Bank energy investments carried out by NGOs in Bulgaria, Cameroon, Colombia, Hungary, India, Indonesia, Kenya, Lithuania, Mexico, the Philippines, Ukraine, Uruguay and Zimbabwe. These NGOs worked together in a collaborative project between 1996 and 1999 to share information and to support each other in relation to the energy sector investments of the World Bank and other multilateral development banks (MDBs) in their respective countries; these NGOs understand the energy situation in their own countries, where they have first-hand experience of the banks' operations in the energy sector.[2] Importantly, the studies produced by all of these NGOs draw the same general conclusion: the way the Bank is implementing its 'reform' programme – which formally started in 1992 and which focuses on privatization of the energy sector – is not leading to sustainable energy.

It is the aim of this report not to launch an ideological attack on privatization, but rather to reflect on its appropriateness in different political environments and different political cultures. Privatization of the energy sector may have positive or negative social and environmental impacts, depending on a range of conditions in the country concerned. Before a country decides whether to privatize its energy sector, appropriate questions to ask are whether the country is a net energy importer or exporter, how many people in the country are connected to an electricity grid, whether the country is primarily industrial or primarily rural and agricultural, and whether the state-owned energy utility has previously been operating at a profit or at a loss. The question of who owns the energy sector is important, but the question of how well the energy sector is governed and managed is more important.

The developed or low-income countries of the world are not being given the opportunity to consider these questions, however. Although it may seem obvious to judge the appropriateness of privatization in relation to the specific circumstances of a particular country, and one would perhaps expect the Bank to be doing so as a matter of course, the World Bank is prescribing privatization for all countries as a panacea, arguing that it will not only encourage more investment into the energy sector, but also improve the performance of energy utilities and reduce energy shortages.

Many of the staff within the Bank understand sustainable energy issues well, and the Bank is responsible for a number of promising sustainable energy initiatives. These staff and initiatives deserve support and encouragement. The Bank has, moreover, produced a substantial amount of high-quality literature on sustainable energy, of which its current energy sector strategy paper, *Fuel for Thought*, is an example. A central problem, which the Bank continues to ignore, however, is the internal institutional barriers to sustainable energy within the Bank itself. There are certainly progressive efforts for sustainable energy being undertaken by a number of divisions and committed individuals within the Bank. Nevertheless, too many of the Bank's staff hold an ideological 'market-fixated' approach to energy development, which is preventing direct support for rural energy, energy efficiency or renewable energy. Taken as a whole the Bank therefore lacks the political will to implement its own policy recommendations relating to sustainable energy and, in practice, the Bank's operations continue to favour business as usual.

Biomass for Survival in Africa[3]

Africa's production of modern fuels is the lowest in the world in spite of its significant energy resources. With the world's lowest per capita consumption of modern energy, increased energy supply through development of indigenous resources (hydro, oil, gas, coal and biomass) should remain on Africa's development agenda. In addition, available fossil fuel resources are not optimally used. For example, in Nigeria, Africa's largest producer and explorer of petroleum, an estimated 8o per cent of the associated natural gas produced is lost though flaring.

With the exception of biomass, renewable energy is at present a minor contributor to the region's energy supply in aggregate terms. However, its potential in light of the decentralized energy needs of Africa's rural population and its environmentally benign character make it an attractive option for meeting future energy needs in the region.

Apart from the relatively industrialized country of South Africa, sub-Saharan Africa has made only a relatively small number of investments in industrial development. The region could ensure that any future industrial investment embodies energy conservation and includes the most appropriate energy efficiency technologies. In African countries, as in other low-income countries around the world, however, the issue of energy efficiency is a lower priority than the issue of energy availability. There are various reasons for the inefficient use of energy in Africa. These include: the transfer from industrialized countries of heavier, less efficient, industries and technologies – including used 'second-hand' products; 'donor-driven' development programmes that favour imports of inappropriate goods and technologies rather than the development of local production facilities and expertise that could provide a basis for efficient energy management; issues related to energy costs; complicated patent laws; inadequate maintenance and operating procedures – often due to lack of affordable spare parts; inadequate national energy research and planning institutions; and a lack of appropriate, effective policy instruments.

Energy planners generally disregard biomass. Nevertheless, it provides 14 per cent of global energy use, and three-quarters of the world's population depend on biomass as their major source of primary energy. Independent reports on energy have long called for national policies and programmes to promote sustainable management of biomass resources for energy and other uses. Recently, major international and national actors have begun to show interest in the development of biomass energy. This is linked to the possibility that a new regime of energy taxes and subsidies – brought in to combat the threat of climate change – will make biomass projects more profitable. In fact biomass is in the process of being re-discovered as ways are being sought to reduce emissions of CO_2 (carbon dioxide), the major greenhouse gas. Biomass is a renewable source of energy that adds no net CO_2 to the atmosphere as long as the amount burned is at least equalled by the amount replanted, and a strategy to replace fossil fuels with biomass may be more cost-effective than planting trees to absorb CO_2 emissions from fossil fuels. A net loss of biomass is taking place in Africa due to its excessive use in the form of wood and charcoal; the middle classes in Africa (teachers, civil servants, small businesses) depend overwhelmingly on charcoal as their main fuel source, followed by fuelwood, kerosene and electricity (when available). There is, further, a contradiction between Africa's dependency on biomass and its enormous reserves of oil, coal and gas. Egypt, Zaire, Ghana and Cameroon have made a strategic choice for hydroelectric power. Nigeria is a net oil exporter while Angola, Chad, Sudan, Congo Brazzaville, Congo Kinshasa and Cameroon are increasing their oil export capacities.

Carbon taxes on the use of fossil fuels may increase the competitive edge of biomass fuels. In Africa, however, many countries, such as Kenya, Ethiopia, Tanzania and Rwanda, spend over 40 per cent of their import budget on oil. The prospect of reducing large oil import bills is, therefore, more likely than the displacement of CO_2 to be an incentive for the development of biomass.

The resurgence of biomass may provide an opportunity for Africa to develop one of its major energy resources. Unfortunately, it is generally ignored by the development banks since it is generally a free good and is therefore not factored into economic models. Certainly the banks have generally failed to provide investment to develop the potential of biomass as a fuel source in the large number of low-income countries, such as those in Africa, where its potential is very large.

Diverse Energy Resources in Asia[4]

Asians in general are more dependent on fossil energy than on biomass. The region has experienced strong economic growth over the last 20 years and until the recent economic downturn, Asia had experienced a period of 'economic re-emergence' characterized by an era of almost uninterrupted high economic growth. This was achieved by transforming the nature of Asian economies from primarily agricultural to primarily industrial and agro-industrial. The high growth era has impacted on the energy demands of the region, and electricity demand in Asia is still expected to grow dramatically in the twenty-first century. The largest countries in the region are China, India and Indonesia and these have seen great increases in industrialization and urbanization, and the growth of a large new middle class, which has created a surge in demand for energy, particularly in the form of electric power. China has experienced an average growth of 9 per cent of gross domestic product during the past two decades, and has both expanded its domestic production and increased its imports of energy, making it the second largest energy consumer and producer in the world, following the United States. Meanwhile, India's capacity to generate electric power has doubled every nine years since 1950 and consumption of electricity is currently growing at around 7 per cent per year.

The expected increase in energy demand in the region has been used as a reason to carry out structural changes in the energy sectors of most

Asian countries. In general, this has taken the shape of a shift towards privatization and commercialization of the energy sector, and energy investments in the region now draw heavily from private sources. It is important to note that this shift towards privatization and commercialization is not being pushed by MDBs alone in Asia, but is also in line with the 'free trade' agenda of powerful regional organizations, such as the Asia–Pacific Economic Cooperation (APEC) forum, for example, which is pushing for competition, removal of subsidies, full cost pricing and more extensive private investment in the public utility sector.

Privatization of state-owned electric utilities and liberalization of the power generation and fossil fuel sub-sectors have become the most noticeable structural reforms that the Asian countries are undertaking in the energy sector. Within this package of reforms, moreover, much attention has been paid to the need for increased energy efficiency.

Energy resources in Asian countries are diverse. Oil reserves, mostly situated in the People's Republic of China and India, are about 46 billion barrels, or around 5 per cent of world reserves. Coal reserves in Asia make up 33 per cent of world reserves, at around 401 billion tons. Natural gas reserves are estimated at around 331 trillion cubic feet. Further, Asia's developing countries have a geothermal power potential of 16 GW and a hydropower potential of 650 GW. There are a number of renewable energy systems in use in the rural areas of Asia. The most widely used are small-scale hydropower, biogas, solar photovoltaic, solar thermal, and wind.[5] It is also important to note that industries in Asia are not highly energy efficient. Power utilities in Asia, for example, have total system losses that range from 25 to 35 per cent;[6] the potential for energy efficiency measures in the region is therefore very large.

Energy Inefficiency in Central and Eastern Europe[7]

The energy sector in CEE countries can be characterized by one word: 'inefficient'. Energy intensity (the amount of energy needed to produce one unit of GDP) is 1.5 to 6.4 times higher than in European Union (EU) countries, and typically twice as high as in most industrialized countries. The disintegration of Communist Party rule in the region in the late 1980s has led to radical changes in the energy sector. Generally, there has been a decline in the consumption of energy in the industrial sector but increased consumption in the commercial and residential sectors. So economic output has fallen faster than energy consumption, causing overall energy intensity in the region to *increase*. This means that the region has become even less energy efficient than it was under Communist Party rule.

This is not to say that the Soviet regime created the conditions for energy efficiency. On the contrary, the power sector in the CEE region under the Soviet-run Council for Mutual Economic Assistance (COMECOM) was notoriously inefficient and contributed substantially to poor environmental quality and especially to high air pollution. Indeed, the large amount of heavy industry, the use of low-quality coal and oil, bad maintenance, and a general lack of awareness of environmental issues during the communist era all contributed to the environmental problems that still play a significant role in the region's energy sector today. Most governments in the region have declared improvements in their air quality. Consistent official data to back up these claims are not available, however, and air quality is measured only in relation to sulphur dioxide (SO_2) emissions. It is further important to realize that most of the pollution reduction in the region to date has been caused by a general slowdown in industrial activity, rather than any 'greening' of energy technologies.

Most CEE countries are limited in their traditional energy sources, and there is high dependence on the import of oil, gas and nuclear fuel from Russia. This is partially because of cheap Russian supply and the regional infrastructure, which was geared for Russian imports. The CEE countries were connected to the Unified Power System (UPS) that comprised all CEE national energy grids. In 1995 Poland, Hungary, the Czech Republic and Slovakia joined the West European Electricity Network, the UCPTE, and other countries in the region were connected later. There are at present insufficient interconnections to allow full and open energy trade to take place in the region. But if current plans are implemented there will soon be full 'freedom of movement' of electricity.

The CEE countries of Bulgaria, the Czech Republic, Estonia, Hungary, Latvia, Lithuania, Poland, Romania, Slovakia and Slovenia have applied for membership of the EU. Enlargement of the EU will be a key political process in Central and Eastern Europe in the coming years, which will influence the region's energy sector. Importantly, EU membership will require that all candidate countries make market and other reforms in their energy, transportation and environment sectors.

Nuclear energy has a special position in most countries of the region. Indeed, for some it is the main source of electricity production. It accounts for 40 per cent of electricity production in Bulgaria, for example, 42 per cent in Hungary, 77 per cent in Lithuania and 49 per cent in Slovakia.[8] In all the countries of the region, there is also a very close relationship between governments and the nuclear industry, either informally, or on a formal basis (for example, through state ownership). All reactors in the region were built – either completely or partially – with major direct or indirect state subsidies.

Nuclear safety standards in CEE countries have been much lower than those in Western Europe and the United States. The International Atomic Energy Agency has commissioned several studies that identify a number of safety problems with nuclear reactors in the region, particularly older versions of pressurized water reactors, and there are calls from citizens' organizations and from government agencies alike to shut these down as soon as possible.

CEE countries in general have a relatively high energy supply capacity. Moreover, all the countries of the region are major per capita emitters of greenhouse gases and have energy intensity rates that are several times those of most other industrialized countries. Integrated resource planning and demand-side management, therefore, have a major role to play in the region (see Box 2.3). Many countries in the region, among them Poland and the Czech Republic, rely heavily on soft brown coal for power generation, which causes particularly noxious air pollution problems. Others, such as Slovakia and Bulgaria, rely heavily on nuclear power, which carries with it the associated problems of waste, safety and nuclear weapons proliferation.

In the long term, the region would benefit from a shift in the fuel mix towards greater use of renewable energy technologies. However, before this can be done, it is imperative that the large inefficiencies in the region's energy systems be addressed. It makes little sense to increase power supply – from whatever source – in a region with such an enormous potential for energy efficiency improvements. Inefficiencies in the CEE energy systems are so large, in fact, that they can be viewed as a major energy resource. Indeed, achievable estimated savings are as high as 20 to 40 per cent, and overall energy savings of 30 per cent could be realistically achieved by the year 2010.

In certain countries of the region, economic pressures are expected to result in a substantial drop in energy use, even without specific measures to reduce fuel consumption. For example, in Ukraine, where imported energy is heavily relied upon, even a 'business as usual' scenario predicts a 34 per cent drop in energy use in response to these economic pressures.[9] Heavy industry has historically played the dominant role in the region's economies. This is changing, however, as regional energy prices approach the level of world prices, causing older, inefficient industries to close down, and others to streamline. The shift from state planning to market economies has already started to cause this to happen, and regionally, energy use (and, consequently, carbon emissions) from the industrial sector have dropped significantly since 1990. On the other hand, consumer demands for items like household appliances and private automobiles are expected to continue to rise, with a resultant rise in CO_2 (the major greenhouse gas causing climate change) emissions from the consumer and transport sectors. Indeed, although total CO_2 emissions have decreased since 1990 (largely because of the failure of inefficient heavy industries), they have already risen in the transport and consumer sectors, and this will continue unless policies are successfully introduced to control fuel consumption in these sectors.

Increasing Urbanization in Latin America[10]

The Latin American energy sector is characterized by marked differences between countries and sub-regions. Such differences are related – among other aspects – to the available energy sources used in each country, to the volume of production and consumption (global and per capita), to the uses of energy and to the level of energy self-sufficiency of each country. Four countries (Argentina, Brazil, Mexico and Venezuela) concentrate more than 80 per cent of the energy produced and consumed, while the majority of Latin American countries have low levels of production and consumption, being marginal in the regional context.

A group of countries produce and export oil, others are net importers, while a third group produces oil but does not have a surplus for export. Some countries produce energy in excess, while others do not have enough and must resort to imports. The main energy sources vary from country to country. In some cases of predominantly rural countries – such as Haiti, Honduras, Guatemala, Guyana and El Salvador – more than 50 per cent of the energy is produced by fuelwood and charcoal, while in other countries, such energy sources represent less than 10 per cent of the energy balance. Only three countries (Argentina, Brazil and Mexico) produce electricity from nuclear reactors.

While in some countries electrification benefits more than 90 per cent of the population (as in the case of Uruguay, Argentina, Brazil, Chile and Venezuela), in others between 20 per cent and 35 per cent of the population do not have access (such as Paraguay and Ecuador), while in Central America access is restricted to less than 50 per cent of the population.

There are also marked differences in relation to the consumption of energy by different economic sectors. In some countries the main consumer is transport, while in others it is industry, and the weight of the residential sector also varies widely from country to country.

Perhaps the only common characteristic of the regional energy sector is that it was almost entirely state-controlled and that the governments' approach was to subordinate market logic to political and social considerations and to consider electricity as a public service and not a profit-driven activity.

Given such heterogeneity – including enormous differences inside different countries in the region – solutions to the region's different problems would need to be tailored to its different realities. However, recent trends clearly show that the same solution is currently being put forward for every single country. Moreover, it seems that this single solution does not respond to the specific needs and realities of each country, but to a programme with an explicit ideology, developed centrally by the international financial organizations and implemented – with minor adaptations to local realities – through the leverage represented by credits.

It is difficult to understand how the same 'solutions' – chiefly privatization, de-monopolization, commercialization and regional integration – can be pursued in countries with an energy surplus as well as in countries with serious external energy dependency. They are also equally implemented in highly electrified countries and in countries where more than half of the population does not have access to electricity; in highly industrialized and urban countries and in mainly agricultural and rural countries.

In truth, what is being implemented in the energy sector is no different from what is being carried out in all the other sectors of activity. Using the argument of the inevitability of globalization, a new economic paradigm is being imposed which, among other things, states that access to national and natural resources must be open to all under market rules. The aim is to downgrade the importance of the state and to assign it a subsidiary role, to liberalize markets and to open up barriers to the entry of foreign capital.

Within this approach, private companies must concentrate on developing their comparative advantages and the state's intervention is limited to guaranteeing free competition and a stable economic, social and political environment that stimulates investments and technical improvements.

Within the energy sector, the aim is to introduce competition, supporting the liberalization of the energy markets and regulating natural monopolies. One of the main aspects of this policy implies the privatization of state enterprises and an increased participation in the energy sector of capital markets.

One of the arguments used for the reform of the energy sector is the need for vast capital to develop the required infrastructure investments to respond to increased demand. The idea is that such investments should be made not by the state but by private enterprises, hence the need to make the energy sector attractive through privatizations and adequate regulatory frameworks. Within this logic, in Latin America regional energy integration has become a major goal and natural gas a major tool to achieve that goal. Indeed, all the above – privatization, integration, promotion of natural gas – are being implemented at great speed throughout Latin America, dramatically changing the situation.

The main – at least theoretically – renewable energy sources used in the region are hydropower, fuelwood/charcoal and sugarcane bagasse (the dry pulp that remains after the juice is extracted). Geothermal energy is important in only two countries (El Salvador and Nicaragua), while wind and solar energy have not yet taken off in the regional context.

Hydropower is a very important energy source in Paraguay (62 per cent of primary energy consumption), Suriname (34 per cent), Costa Rica (26 per cent), Uruguay (20 per cent), Brazil (18 per cent), Honduras (13 per cent) and Peru (11 per cent). During the 1970–95 period, the participation of this energy source increased – in many cases dramatically – in most of the countries in the region and fell only in Bolivia, (from 7.9 to 6.3), Mexico (2.7 to 1.1), Nicaragua (3.3 to 2.0), Dominican Republic (5.9 to 4.8) and Suriname (64.0 to 33.7). In the past hydropower has received strong support from MDBs, which centred their loans on large dams, which have had devastating social and environmental impacts in the region, as elsewhere in the world. Regional opposition to such projects has increased, however, and the banks are now withdrawing from funding large dams. Small-scale hydropower, meanwhile, has received little attention, although the region's hydraulic potential is enormous and the social and ecological impacts of such use would be minimal or largely beneficial.

Fuelwood (including charcoal) is still an important energy source in many countries, with the exceptions of Argentina (1.2 per cent), Cuba (2.4 per cent), Mexico (3.5 per cent), Bolivia (7.4 per cent) and Suriname (9.6 per cent). With few exceptions, the use of fuelwood does not constitute a major cause of deforestation in the region and the incorporation of adequate forest management could ensure the renewable use of a still abundant resource.

Although there is some experience in wood gasification to produce secondary energy, much more research is needed for a more efficient and comfortable use of this traditional fuel. Even when fuelwood is used mostly at the household level, some countries (particularly Brazil and Uruguay) have made ample use of wood as an energy source for industry (charcoal in the former and wood gasification in the latter).

Use of fuelwood has dropped in many countries, having been substituted either by electricity or fossil fuels, particularly liquid petroleum gasoline (LPG) and kerosene. In many cases this has been as a result of government policies. Wood use in Brazil dropped from 64.2 per cent in 1970 to 21 per cent in 1995, in the Dominican Republic from 51.5 per cent to 31.7 per cent, in Costa Rica from 43.1 per cent to 20.4 per cent, in Ecuador from 39.4 per cent to 10.7 per cent, in El Salvador from 79.1 per cent to 46.2 per cent, in Paraguay from 80.1 per cent to 25.8 per cent and in Colombia from 23 per cent to 13.4 per cent. Although in some cases fuelwood has been substituted by hydroelectricity (particularly in Paraguay), in most cases it has meant an increase in fossil fuel use with, of course, a consequent impact on the global environment in terms of the release of CO_2.

Sugarcane bagasse is the third widespread renewable energy source in Latin America. The energy balance of the majority of the countries in the region includes this source as an important component, though only some surpass 10 per cent (Guyana 44.8 per cent, Cuba 44.4 per cent, Barbados 40.8 per cent, Brazil 19.5 per cent, Jamaica 15.6 per cent). In spite of its obvious advantages from the perspective of the global environment, it is not without local environmental problems, particularly linked to impacts derived from extensive 'Green Revolution'-type monocultures. The approach is, however, interesting, in the sense that countries take advantage of locally produced renewable production and transform its 'wastes' into different forms of energy. This approach could be extended to produce energy (particularly modern forms of energy) from biomass produced as a by-product of other agricultural products.

To date, few countries in Latin America have given much attention to geothermal, solar or wind energy, although in most countries in the region there is sufficient knowledge and experience, particularly concerning the latter two.

Energy efficiency has received scarce attention within the region. This is despite the fact that available data indicate that between 10 per cent and 20 per cent of current energy consumption could be saved in the short and medium terms through more efficient energy use. Indeed, not only have there been few advances concerning energy efficiency but, on the contrary, during the 1980–95 period, a decrease in energy intensity (the amount of energy needed to produce a unit of GDP) took place in the region as a whole. In fact only three countries improved their situation with regard to energy efficiency (Guyana, Jamaica and Uruguay), while six others achieved no improvement and the remaining 13 worsened their situation.

This decrease has been caused by an accelerated urbanization of Latin American societies and the growing consumerism of its population. The increased concentration of the region's population in large cities is a crucial factor since it has increased energy requirements, in both the residential and transport sectors. This has been coupled with an intensified use of electrical appliances and an increase in the number of private cars (at the expense of public transport systems), implying a growing consumption of fossil fuels for motor vehicles, which are the main consumers of hydrocarbons and the main source of atmospheric pollution within the region's energy sector.

A World of Difference

Expected global growth in energy demand is staggering: more electricity-generating capacity will probably have been built over the period

1990 to 2020 than was built in the previous 100 years. A 55 per cent increase in global energy consumption, which will take place mostly in low-income countries, is expected between 1998 and 2020.

The problem facing the industrialized nations in relation to energy is mainly one of excessive levels of demand. The industrialized countries support extravagant levels of material consumption by providing food, water and energy through wasteful, centralized systems that depend on high inputs of energy, capital and materials. The price for this extravagance is a deteriorating environment within the industrialized countries themselves, coupled with the global threat of climate change.

The problems facing low-income countries in relation to energy are diverse and complex. First, energy is an essential input to the process of production, and here low-income countries face two main problems: fuel costs and continuity of energy supplies. These problems – which can be considered as a crisis of energy for economic development – determine both the pace and the type of economic development that is possible in low-income countries. Second, energy is a basic need, which is essential for survival, and the ability of many communities to provide energy for basic needs is threatened. Unfortunately, this crisis of energy for survival, which is affecting the poorest people and the poorest countries, is receiving very little attention.

Almost all countries are highly dependent on oil for their commercial fuel supplies. Three-quarters of all low-income countries were oil importers in 1987, and of the 38 poorest countries, 29 imported more than 70 per cent of their commercial energy, which was nearly all in the form of oil. In sub-Saharan Africa, moreover, most countries spent 25–50 per cent of their foreign currency earnings on oil imports.

The risk of economic dislocation associated with overdependence on imported oil became clear after the 1973 oil crisis, when the international market price of 'Saudi light' oil, for example, rose from $3.01 per barrel in October 1973 to $11.65 per barrel in January 1974. Four years later, the price of Saudi light rose from $13.34 in January 1979 to $18.00 in October 1979, $26.00 in January 1980, and $32.00 in January 1981.

Low-income countries were hardest hit by the 1981 price rise. It was accompanied by general economic recession and falling commodity prices. So export earnings declined at the time when they were most needed to cover oil imports. This meant that many low-income countries had problems maintaining their existing levels of oil imports, and the poorest countries – particularly those in Africa and Asia – were unable to increase oil imports, were unable to fuel growth in vital sectors such as transport and industry.

On the positive side, the oil crisis encouraged many countries to initiate

include hydropower

or increase research and development into new and renewable energy technologies, including ways to improve end-use energy efficiency. The inability of low-income countries to influence the cost of energy and commodity prices, and their dependence on fluctuations in international market prices in general, however, meant that they became, in many cases, literally unable to power the conventional process of economic development. The weak bargaining position of low-income countries within the international economy is therefore an essential element in the crisis of energy for development.

Energy is used in daily life for heating, lighting and cooking. Energy can therefore be considered a basic need comparable to food, water and shelter. In many rural areas, people have begun to depend on LPG and kerosene, which have been encouraged by government subsidies. These subsidies have tended to be removed in recent years, however, and rural populations have tended to return to the use of wood, charcoal, and crop and animal residues. Significant numbers of people in low-income countries, therefore, face increasing problems in obtaining energy to take them beyond a level of subsistence. The heaviest share of this burden falls on the least powerful, particularly women, the land-poor, the landless, and the marginalized urban dwellers. The more powerful local groups, meanwhile, tend to be the last to be affected by energy scarcities.

For many people in low-income countries, the main source of energy is biomass, or traditional fuels. This is particularly so in the case of cooking, which is the largest single energy user in low-income country households. Development agencies have ignored the problem of energy scarcity for such a long time that its scale is not fully known. But this lack of knowledge about the problem is also due to the way in which the problem manifests itself. People rarely run completely out of fuel. Rather, they shift to using fuels of inferior quality, changing from wood to stalks or dung, for example, or fuels that take longer to gather, or have to be paid for, where they were previously gathered freely. These difficulties affect a class of people who are economically and politically invisible, so they tend to be ignored by national governments and powerful commercial interests. Moreover, much of the scarcity of traditional fuels in low-income countries is caused by limited access to these fuels, rather than actual physical shortages: the problem is not so much a lack of resources, but rather how resources are controlled.

The problems related to the supply and uses of energy are diverse. Every country has its own unique set of problems and opportunities. So any global energy policy, such as that of the World Bank, should take account of the varying circumstances of different parts of the world, and the particular circumstances of each individual country. The fact that the

World Bank is not doing this is the subject of the following chapter. It describes the Bank's 'one size fits all' reform programme, and explains that the way the Bank is implementing this programme – which focuses mainly on increasing private ownership of the energy sector – is not leading to any significant shift towards sustainable energy.

Notes

1. For example, World Bank official responses to: Environmental Defense Fund and Natural Resources Defense Council, *Power Failure: A Review of the World Bank's Implementation of its New Energy Policy*, Washington, DC, March 1994; B. Rich, *Mortgaging the Earth*, Boston, MD: Beacon Press, 1994; Sustainable Energy and Economy Network (Institute for Policy Studies, USA), International Trade Information Service (USA), Halifax Initiative (Canada), Reform the World Bank Campaign (Italy), *The World Bank and the G-7: Charging the Earth's Climate for Business: An Analysis of the World Bank Fossil Fuel Project Lending since the 1992 Earth Summit*, June 1997.

2. The MDB-Energy Project was implemented by Both ENDS, Damrak 28–30, 1012 LJ Amsterdam, Netherlands. The work is continuing as the TNI Energy Project, based at the Transnational Institute, Paulus Potterstraat, 20, 1071 DA, Amsterdam (see the website: www.tni.org).

3. This section is based on material provided by Econews Africa, Nairobi, Kenya.

4. This section is based on material provided by the Legal Rights and Natural Resources Center, Manila, Philippines.

5. Asian Development Bank Policy for the Energy Sector.

6. Ibid.

7. This section is based on material provided by the CEE Bankwatch Network, Prague, Czech Republic.

8. Figures for 1994, *Transition Report 1996*, European Bank for Reconstruction and Development, p. 38.

9. Ibid., p. 34.

10. This section is based on material provided by the Third World Institute, Montevideo, Uruguay.

The World Bank and Energy

The Shift in the Bank's Energy Policies

During the 1970s and early 1980s the Bank's lending policy for the energy sector focused mainly on developing state-owned monopolies and helping governments to raise financing, both locally and internationally, for power investments. The Bank's policy was that economies of scale make it more attractive for large producers to dominate power generation, and that this ensures that consumers receive electricity at the lowest cost. The Bank also believed that government ownership and regulation ensured that generating capacity was available when needed, and that an appropriate level of investment takes place. In general, the Bank believed that public ownership of energy utilities, and cross-subsidies within the energy sector, in order to meet public policy goals of universal service and affordable services for the less well off, were the most sensible ways forward.

Bank and other studies produced in the early 1980s, however, came to the conclusion that funding requirements for the investment plans of the power sub-sector were much greater than the amounts available either from electric power utility profits or from official financing agencies. The Bank concluded that there would therefore not be enough money available through public funds to finance the expected growth in energy demand in developing countries. On the basis of these arguments a shift in rhetoric took place within the Bank in the early 1980s. Audits and studies carried out by the Bank during the late 1980s and early 1990s further concluded that the poor performance of the energy sector was being caused by a number of additional factors. Among these were: the reluctance of governments to increase the price of electricity; government (political) influence over electric utilities' financial and management decisions; and government attempts to meet socio-political objectives by providing subsidies to various electricity consumers. In other words, Bank studies concluded that the electric power sector was not as economically efficient as it could be because governments were using it to meet overall development objectives, rather than running it purely along business lines.

By the early 1990s internal Bank discussion papers began directly to challenge state ownership of the energy sector. A discussion paper produced in 1991, for example, by the International Finance Corporation – the Bank's private lending arm – stated:

> Intuitively, it is not at all clear why the generation of power should be considered to be a natural monopoly. Unlike distribution, which entails a single network for delivering power to individual consumers, there is no obvious reason why independent producers should not be able to sell electricity to whomever distributes it to the public. Provided tariffs are sufficiently regulated so that monopoly profits are not received, no harm is done in having electricity produced in the private sector.[1]

Like all public policy, this line of thinking within the Bank developed within a political context. In fact these ideas – in relation to international relations and foreign investment – did not originate within the Bank at all, but were developed in a number of conservative 'think-tanks' and university institutes that were formed in the 1970s. Examples of such think-tanks are the London-based Adam Smith Institute and the Washington, DC-based Heritage Foundation, which promote the belief that free market forms of social organization are inherently superior to any kind of welfare state social democracy. During the Reagan/Thatcher era the influence of these and other business–oriented policy development institutions grew. An ideological movement developed, and began to implement what is now an organized international lobby campaign, based on the principles embraced by such institutions. Previously considered radically ideological and 'off the wall', these ideas began to take centre-stage in policy dialogue and policy development in the United States in the early 1980s. In his book on privatization and public sector reform *In the Public Interest?*, Brendan Martin documents how the ideas generated by these and similar conservative pressure groups were first adopted by the United States Agency for International Development (USAID) in a sharp change in policy immediately after the Reagan administration took office in 1981; Martin further shows how the US government and USAID then exerted various forms of influence on the World Bank and other bilateral aid agencies and multilateral development banks to adopt an increasingly free-market approach to development.[2]

In relation to the energy sector activities of the World Bank, this political shift began in the 1980s with the commencement of loans for 'sectoral adjustment' or 'structural adjustment', which were intended to restrain government intervention and spending, and culminated in a new 'reform agenda' contained in two key policy papers approved by the Bank's Board of Executive Directors in 1992. The first of these policy papers was *The*

World Bank's Role in the Electric Power Sector, which became known as the *Power Paper*. It lays out the Bank's plans to move away from supporting 'the single national electric utility operating as a public monopoly [in developing countries]', and to 'aggressively pursue the commercialization and corporatization of, and private sector participation in, developing-country power sectors'.[3] In practical terms, the Bank's energy policies translate this into the splitting up or 'unbundling' of previously state-run energy utilities into separate generation, transmission and distribution companies, which are then privatized, and which must operate commercially in order to gain access to finance under competitive conditions in the global capital markets. At the same time, the rules governing the supply of energy are liberalized, allowing independent power producers (IPPs) to produce and sell energy privately.[4]

From the late 1970s onwards the general public in most countries became increasingly concerned with environmental issues. Furthermore, starting in the early 1980s, NGOs working on environment and development issues became increasingly engaged in criticizing the negative social and ecological impacts of projects supported by the World Bank and other MDBs. By the time the early drafts of the Bank's *Power Paper* were produced in 1991, one year before the United Nations Earth Summit, environmental problems were being placed at the top of the political agenda for the first time in history and were being discussed by leaders of both industrialized and low-income countries. Inevitably then, NGOs and others criticized the *Power Paper* drafts for their lack of attention to green issues. In response to these criticisms the Bank, sensitive to the growing political importance of environmental issues, produced an accompanying document to deal with environmental concerns. This second policy paper was called *Energy Efficiency and Conservation in the Developing World*, and became known as the *Efficiency Paper*. It recommends that energy efficiency and renewables should be better integrated into the Bank's general country policy dialogue with developing countries, that the Bank should give 'high-level, in country visibility' to energy efficiency and demand-side management, and that the Bank should give 'greater attention to the transfer of more energy efficient and pollution reducing technologies in its sector and project work'.[5]

The Policies are Controversial

The Bank's energy policies were written in the early 1990s and reflect the social and political forces that were acting on the Bank at that time. On the one hand the swing to market-based solutions to development problems is expressed in the *Power Paper*. On the other hand a growing concern for the environment is expressed in the *Efficiency Paper*. The main substance of the policies is the ideological element: the commercialization and corporatization of, and private sector participation in, developing country power sectors. This is then glossed over with the environmentalist element: increased attention for renewable energy and energy efficiency technologies. These two elements are not part of a cohesive analysis, though: the policies do not sufficiently demonstrate how energy sector privatization will lead to benefits for the environment. It is not surprising, then, that in practice the policies have largely failed to deliver environmental benefits. This fact has been hammered home in a series of NGO reports since the policies were produced (see below). Furthermore, an internal review completed by the Bank's Operations Evaluation Department (OED) in February 1998 revealed that the Bank has had little success with its reform agenda and is not sure to what extent its energy policies will be able to lead to environmental benefits.[6]

One of the Bank's central arguments, for example, is that privatization of the energy sector leads to increased end-use energy efficiency. The Bank argues that privatization removes cross-subsidies for energy, which leads to increased energy prices, which in turn encourages consumers to use energy more efficiently. Even this central assumption, however, is brought into question by the OED's February 1998 review of Bank energy lending, which admits that:

> Little information is available on the impact of increased average tariffs on consumption in countries where the power sector has been privatized … In some cases, removing cross-subsidies may actually result in increased consumption for certain consumers if their tariffs are lower than before privatization. This is particularly true for industrial consumers – usually the largest users of electricity in developing countries – because they have been cross-subsidizing residential and agricultural consumption in many countries. In Argentina, for example, post privatization comparisons show that prices for all consumer categories (except for the low-consumption residential group) declined by an average of 10 percent.[7]

This uncertainty shows how one of the central tenets of the Bank's reform process is not based on any conclusive evidence.

The OED report further revealed that where privatization has occurred there is a trend towards power plants with low investment costs, particularly coal-fired plants, because these are relatively cheap to install. A Bank survey of investments between 1991 and 1996 showed that one-third of the generation was coal-fired and that the trend seems to favour more coal. In Asia half the privately financed generation capacity was coal-fired. Even in Latin America, which has traditionally based its power generation on hydroelectric dams, there were agreements with independent power producers for twice as much coal-fired power as for hydroelectric power.

Some energy analysts even doubt that, in the long term, the Bank's reform process will lead to an increase in economic efficiency and a drop in energy costs. They argue that the monopolistic nature of the energy sector has, until now, resulted in high costs of labour and relatively low costs of capital. In a liberalized, privatized industry, they say, labour costs will probably decrease, but capital costs will grow. In the end, the increase in capital costs is likely to outweigh the lower labour costs, leading to relatively higher electricity costs in the long run.[8]

Today, seven years after their adoption, it is clear that the Bank's energy policies have been extremely difficult for the Bank to implement. There is general support within the Bank – for example among country managers and task managers – for privatization of the energy sector; there has been an internal effort within the Bank to educate staff about energy sector privatization and how to carry it out. Nevertheless, progress has been slow. One of the major reasons for this has been what the Bank calls 'weak government commitment'. This is another way of saying that borrowing governments have been resisting the reforms proposed by the Bank. In fact, many borrowing governments share the analysis of many of the Bank's critics, which is that privatization of the energy sector amounts to externally imposed measures designed to benefit foreign investors. The Bank's response to such dissent has been aggressive. In the first half of the 1990s, Kenya suffered a freeze on loans from the Bank, partly due to the Kenyan government's refusal to implement energy sector reforms related to private-sector involvement in the power sector. As a result, investments in generating capacity were halted, causing power outages, load shedding, rationings and blackouts. In general, the Bank has tended to characterize those who dissent from its privatization agenda as being politically or ideologically motivated. This is illustrated by an internal project appraisal report for a 1999 Bank loan to restructure the energy sector in the Indian state of Andhra Pradesh, which states that:

> The reform programme is likely to continue to face strong opposition, as it will impinge on large and powerful vested interests. The political opposition and vested interest groups have used and will continue to use measures like privatization of distribution and tariff adjustments as points of contention.[9]

The Bank's February 1998 internal review of energy lending clearly spells out its difficulties in implementing its privatization programme:

> Few completed projects have included privatization components (Honduras, Morocco, Pakistan, India). [Moreover,] objectives for these few projects were only partly achieved.

For example, the Energy Sector Loan ... for Honduras helped introduce a new Electricity Law, but the law had some deficiencies. While there has been private sector financing of generation and a contract plan for the state-owned power company, the government has retained political control and made it difficult for the company to operate as a fully autonomous entity.

An Electricity Law was also drafted in Morocco ... but government approval was still pending at completion of the project. In Pakistan ... most policy objectives were only achieved during subsequent operations and there were important shortcomings in institutional development of the major power utility. In India ... a clear policy framework at the state level and lack of creditworthy clients inhibited much progress on privatization. Only four state electricity boards used the proceeds of the loan and as a result two independent power producers (IPPs) started producing electricity ...

In many cases some of the more difficult decisions still have to be made, particularly those relating to price reform and privatization of existing facilities ... Countries such as Ecuador, Sri Lanka, Ghana, Guinea, and Bangladesh have been wavering in their commitment to restructuring the power sub sector and making tariff adjustments ...

Only about 15 countries have started to sell off existing generation capacity to the private sector. Practically all transmission and distribution (T&D) is still in the hands of the public sector (except in Argentina, Chile, Peru, Bolivia, and Hungary; Brazil is just starting) ...

Privatization of existing assets, particularly distribution, has barely begun ... Clearly then, implementation of the reform agenda will take considerable time.

The appropriateness of privatization depends on the specific conditions in the country concerned. With proper regulation it may offer opportunities for small independent producers of renewable energy to replace monopoly power utilities, and can also lead to energy savings on the demand side. Indeed, privatization gives an opportunity for the establishment of new government regulations that encourage sustainable energy and extension of energy services into rural areas. Energy investments can be required to incorporate integrated resource planning, for example, making full use of demand-side resources (see Box 2.3).

Rather than judging the appropriateness of privatization in the case of each of its client countries, however, the Bank is prescribing the privatization agenda as if 'one size fits all'. Privatization is being advanced by the World Bank as a panacea that will improve the technical and managerial performance of energy utilities, encourage more investment in the energy sector, and reduce energy shortages. This is regardless of whether the borrowing country is energy-dependent or an energy exporter, regardless of how many people in the client country are connected to an electricity grid, regardless of whether the client country is primarily industrial or primarily rural and agricultural, and even regardless of whether state-owned energy utilities have previously been operating at a profit or at a loss.

Box 2.1 Local communities fight for public power in the USA

The pre-eminent free-market economy is, of course, the United States where, today, about 260 privately owned power utilities supply 75 per cent of US customers, 940 rural electric cooperatives serve 10 per cent of US customers, and 2000 public power systems serve 14 per cent of US customers.[10] Publicly owned energy utilities sell power to customers at cost price and their rates are on average 33 per cent lower than publicly owned utilities. Their services are as good as or better than privately owned utilities, and their revenues are reinvested in the local economy. Nevertheless, privately owned utilities still dominate the US energy sector. In his book *The Power Makers*[11] Richard Munson documents how privately owned utilities in the USA have maintained their superior position through brilliant propaganda campaigns and the capacity to buy political influence. Indeed, after one advertising campaign during the McCarthy era in the 1950s, in which public power was equated with socialism and loss of freedom, support for public power dropped from 70 per cent to just 30 per cent. At that time the private energy industries also altered their terminology: from privately owned utilities to their current title: investor-owned utilities, or IOUs. Today, there are examples of local communities in the USA, such as the people in the townships and villages in St. Lawrence County and adjoining Franklin County in northern New York State, who are organizing to reverse the trend towards privatization of the energy sector and to establish publicly owned 'municipal power' companies. They argue that this will both save money and help create local employment.[12] On the west coast of the USA, meanwhile, in northern California, cities, counties and community groups are struggling to form publicly owned cooperatives in order to break what they see as the 'stranglehold' of the investor-owned Pacific Gas and Electric Company. Their aim is 'to reduce the astronomical rates we pay for electricity, and bring vast amounts of money into public coffers'.

Box 2.2 Controversy surrounds UK energy privatization

The Bank's reform agenda takes, as one of its main examples, the energy sector liberalization that took place in 1990 in the UK. The 'British electricity experiment', in which the Central Electricity Generating Board was restructured and privatized, has been heavily criticized, however, mainly because of the way large profits brought about by liberalization have been divided. In Britain, there is a widely held view that the profits created by increased economic efficiency in the sector have not been passed on fairly to consumers but have instead been spent on new high-risk ventures and higher payments to shareholders.[13]

Privatization *per se* is not the issue here. The concern is rather the dogmatic way in which the World Bank is seeking to implement privatization as an objective in itself, rather than as a means to promote sustainable energy.

The Bank's dogmatic approach to privatization can be demonstrated by, for example, the way in which it deals with integrated resource planning (IRP) and demand-side management (DSM) for the energy sector (see Box 2.3).

The World Bank's 1992 energy policies devote several pages to underlining the importance of these policy instruments for increasing energy efficiency. Official Bank figures for lending for energy efficiency over the past four years, however, comprise only about 1.6 per cent of total energy sector lending.[14] In fact, the actual amount lent on direct measures to increase energy efficiency is likely to be much less. An example of this is the Bank's ongoing Energy Sector Reform and Power Development Project in Kenya. This project includes a component for energy efficiency. But on closer inspection a substantial amount of the loan is being spent on improvements on the *supply* side, such as line loss reduction. Furthermore, commercialization of the energy sector is listed in the loan as a measure to increase general efficiency. Meanwhile, energy conservation, which is a crucial aspect of integrated energy strategies, is not emphasized within the project.

Indeed, most recently, Bank reports have even started to question the efficacy of DSM as a tool for energy planning.[15] Meanwhile, Bank staff and consultants are tending to complain that they are not able to identify sufficient numbers of good DSM projects.

According to the Indian NGO Prayas, this problem is caused in part by the Bank's 'market-locked thinking'.[16] The World Bank believes strongly in market mechanisms and expects private investors to take up sustainable energy projects. Private investors, however, have little experience with sustainable energy, which they view as 'high risk'. Power utilities – whether publicly or privately owned – could encourage investment in sustainable energy by proving that such projects can be profitably carried out. In most countries, however, power utilities are showing little interest. The piloting and demonstration of sustainable energy projects could help to break this vicious circle. But in the case of India, taken as just one example, power utilities are not willing to allow DSM pilot or demonstration projects, even on a shared profits basis. The World Bank argues, correctly, that this problem is due to the non-commercial nature of the Indian power sector. But then the Bank goes on to argue, without any supporting evidence, that

---

Box 2.3 Integrated resource planning and demand-side management

Integrated resource planning (IRP) is a process that has been used successfully to date by energy utilities in the USA, Canada and Australia. IRP considers both the supply side (such as building a power plant) and the demand side (such as replacing traditional incandescent lights with more efficient fluorescent lamps). IRP sees demand-side energy savings as a legitimate substitute for creating more energy supply. IRP usually produces an analysis that ranks the available supply- and demand-side options, taking into account environmental considerations, reliability and availability.

IRP is not an alternative to building new power plants. Rather, it is a way to rank all the available demand-side and supply-side 'resources' in terms of how much money they will cost, the environmental damage they will cause, and the social impact they will have. All energy options cost money and have social and environmental impacts. IRP allows all of these 'costs' to be taken into account, and shows which is the 'cheapest' option (for this reason IRP is also sometimes referred to as 'least-cost' planning).

IRP is potentially applicable to any electric utility, large or small, public or private.

Within IRP, schemes that reduce energy demand – such as efficient lighting or motors, or better insulated housing – are known as demand-side management, or DSM.

full commercialization (full privatization) of the power sector will solve the problem.

The adoption of market principles can in many cases be an impetus for sustainable energy, while state control of the energy sector, such as in India, can be seen as an impediment to the development of new and renewable energy technologies. There are problems, however, in the Bank's 'market-locked' approach to the problem. First, as described earlier, the Bank's aggressive efforts to privatize the energy sector are not appreciated by most developing country governments and are proceeding with great difficulty.

Who knows how long it will take before full privatization is achieved, if ever? In the meantime, huge opportunities are being wasted in obtaining field experience in sustainable energy projects. Second, the evidence to date shows that private investors are more interested in large-scale power generation projects than in the relatively cumbersome contracts and monitoring involved in smaller-scale renewable energy and energy efficiency investments. It is obvious, therefore, that more needs to be done by the Bank to promote sustainable energy than simply to evoke the power of the market.

Another area that demonstrates the Bank's obsession with the market is that of rural energy. After being criticized for not paying enough attention to rural energy issues, the Bank produced a draft rural energy policy in 1995 and held consultations with donor agencies and others on this document. The draft was finally produced as a book in 1996, called *Rural Energy and Development*, which was subsequently given the status of a 'Good Practice' paper (see the explanation of Good Practice, below).[17] This means, however, that although the book may influence discussions within the Bank it is not mandatory policy and Bank staff are not required to read it or to implement any of its recommendations. The book gives a good analysis of the rural energy situation, options for rural electrification, and innovations in rural energy. Differences of opinion within the Bank are clear, however, from the internal contradictions within the book's analysis. At one point, for example, the book raises doubts about the overall privatization agenda:

> Liberalizing energy markets, however important, may not be the complete answer. Despite the progress made in encouraging private investment in the electricity industry since the beginning of the 1990s, for example, private companies have shown little interest in extending electricity supplies to rural areas. They have instead preferred to concentrate on more lucrative contracts to generate electricity and to supply industrial and urban customers. There is evidence, in other words, that creating urban-based energy markets by itself will fail to provide rural electricity.

The opinion of senior Bank management, however, expressed in the book's foreword by the vice-president of finance and private sector development, Jean-François Rischard, clearly reasserts that the Bank's primary task is that of political reform:

> In recent years the Bank's work on energy has largely focused on making existing energy supply and consuming industries more [economically] efficient, opening them up to competition, and encouraging private sector participation. This is an important job and far from finished ... Rural energy presents the Bank with an important challenge. It should be a key part of our work. [But] ... while the Bank can play an important catalytic role, *real progress in tackling these issues is critically linked to the adoption of the needed policy and institutional reforms.* [emphasis added]

When the Bank refers to 'the needed policy and institutional reforms' it means, of course, privatization of the energy sector. The Bank is clearly revealing an ideological 'market-fixated' approach to the problem here: even faced with evidence that privatization alone is not leading to any progress on rural energy, senior Bank staff still insist that there is no other way to promote rural energy than to privatize the energy sector.

It would be wrong to suggest that the Bank has done nothing to promote sustainable energy. In fact there are groups and individuals within the Bank who are making strong efforts to push the institution in the direction of sustainability, and the Bank has taken many initiatives that deserve support. Since 1983, the Bank has run an Energy Sector Management Assistance Program (ESMAP), which provides technical support for the implementation of energy efficiency technologies. It has a programme that, since 1989, has aimed to support the development of small-scale power options that complement electricity supplies to the grid, called Financing Energy Services for Small-Scale Energy Users (FINESSE). The Bank also has an Asian Alternative Energy Unit (ASTAE), which is promoting renewable energy projects in Indonesia, India, Sri Lanka and China. The Bank houses the Global Environment Facility (GEF), which was created as a result of the 1992 Earth Summit and which is intended to cover the problems of climate change, ozone depletion, marine pollution and biodiversity. In relation to energy the GEF is supposed to help low-income countries to bear the extra costs of measures that mitigate global environmental effects (mainly CO_2 emissions, which are the single largest source of global warming). The Bank's private sector lending arm, the International Finance Corporation (IFC) has launched a Photovoltaic Market Transformation Initiative (PVMTI), to award competitively $30 million, through the GEF, to companies promoting solar PV. The International Finance Corporation has also created a commercial debt and equity fund, capitalized at between

$150 and $240 million, called the Renewable Energy and Energy Efficiency fund (REEF) to promote small renewable and energy efficiency projects. The Bank initiated an Activities Implemented Jointly Program (AIJ), with finance from the Norwegian government, to investigate the potential for 'joint implementation' of government commitments to reduce greenhouse gas emissions. Further, the Bank, the IFC, US-based private foundations and others have set up a Solar Development Corporation (SDC) to provide business development services to local solar entrepreneurs, and to provide credit for both solar businesses and purchasers of solar home systems. The Bank and several cooperating institutions also completed a Carbon Back-casting Study to demonstrate how the recent energy portfolio of the Bank would have been impacted by the incorporation of a 'shadow value' for carbon emissions. The Bank has a Clean Coal Initiative, which is intended to encourage the use of so called 'environmentally friendly' coal techno-logies. It further has a Regional Program on the Traditional Energy Sector (RPTES) operating in Africa, to study the functioning of traditional, informal, energy markets. Most recently the Bank has started to develop plans for a programme of Energy–Environment Sector Work (EESW), which will 'help map out what the Bank Group will do'. The Bank has estimated the cost of the EESW programme at around $2 million per year and has discussed financing of the programme with ESMAP donors and others such as bilateral donors and the GEF. Further, the Bank runs 'EMPower Info', which provides a broad range of information about the environmental assessment process, various energy sources and their en-vironmental impacts, and power conversion and pollution mitigation technologies and standards. This aims to help policy-makers to incorporate environmental issues into planning power projects, evaluating alternative power sector strategies, and setting environmental standards. The Bank has also launched a Prototype Carbon Fund (PCF) to 'facilitate exchanges of emission reduction units under the provisions in the Kyoto protocol' (see Box 2.4).

A number of these initiatives should be applauded. Others, such as the GEF and the PCF, for example, have become controversial for a variety of reasons and need critical attention. A general problem that all these initi-atives face, however, is that they do little or nothing to address mainstream Bank lending, which is still overwhelmingly focused on supporting business as usual, rather than sustainable energy. There are various reasons for this general problem. To begin with, progressive initiatives are very small compared to the Bank's overall portfolio and are generally dependent upon external funding from developed country governments, rather than being wholly financially supported by the Bank. The fact that many of the Bank's progressive initiatives for sustainable energy are largely externally funded means that they tend to be poorly institutionalized within the Bank and have, therefore, still a limited impact: they tend to be 'sealed off' from traditional bank operations and are not taken as seriously as mainstream activities by senior managers, or by operational staff. All of the initiatives mentioned had strong political and financial support from outside the Bank, mainly from shareholder governments that have taken a 'progressive' position in relation to sustainable energy, such as Denmark, The Netherlands and Sweden. It is not likely that any of the initiatives mentioned above would have emerged without strong political and financial support from outside the Bank.

There are also institutional factors within the Bank that tend to work against sustainable energy. To begin with, the sheer size of the Bank creates a problem; the Bank's overhead costs are high so there is a tendency to make investments large. It takes the same amount of staff time for the Bank to process a loan for several million dollars for a coal-fired power station, for example, as it does for a larger number of more complicated energy efficiency projects, which will not produce the same high return for the Bank in terms of overheads. This method of working does not match the needs of smaller-scale new and renewable energy technologies, which typically need less in terms of cash and more in terms of technical assistance.

The way the Bank calculates the costs and benefits of projects also works against new and renewable energy technologies. The Bank does not generally take full life-cycle costs into account in procurement decisions, and this puts energy efficiency technologies in particular at a disadvantage, since these tend to be relatively expensive to install, but save money over their lifetime. In fact, many Bank staff are simply unaware of the potentials of energy efficiency and renewables. Energy efficiency, for example, is viewed by many Bank staff as irrelevant to the economies of low-income countries. In addition, annual lending targets put pressure on loan officers in the Bank to make large loans, rather then the smaller and less easily managed loans that characterize energy efficiency and renewable energy projects. Loan officers, further, face restrictions in the grants they are allowed to make for technical assistance; this makes it difficult for loan officers to provide assistance for energy efficiency, which typically requires a relatively large amount of technical assistance. The most recent attempt to address these problems has been to create a Thematic Group on energy with a mandate to promote the sustainable energy agenda from within the Bank's Energy, Mining and Telecommunications sector. At the time of writing, however, it is too early to judge whether this strategy will dislodge the strong vested interests promoting business as usual within the Bank.

Deregulation, commercialization and privatization of the energy sector are not necessarily better or worse for sustainable energy. The crucial factor is the *way* in which privatization takes place. In order for any reforms to promote sustainable energy, however, whether or not privatization takes place, effective regulation of the energy sector is crucial. The Bank's 1992 energy policies are very clear on the need for a transparent regulatory process for the energy sector. The *Power Paper* states that:

> A requirement for all power lending will be explicit country movement toward the establishment of a legal framework and regulatory processes satisfactory to the Bank. To this end, in conjunction with other economy-wide initiatives, this requires countries to set up transparent regulatory processes that are clearly independent of power suppliers and that avoid government interference in day-to-day power company operations (whether the company is privately or publicly owned). The regulatory framework should establish a sound basis for open discussion of power-sector economic, financial, environmental, and service policies. The Bank must be satisfied that there is meaningful progress towards this objective.

The Bank, however, has not followed this up whole-heartedly and examples of regulatory failures in borrowing countries are legion. Indeed, the regulatory process has tended to be hijacked in most countries by narrow and strong vested interests that prevent open and transparent discussion of development objectives, and any meaningful shift towards sustainable energy. The Bank is culpable insofar as it has not taken this problem seriously enough. In fact, despite the clearly defined requirement in its energy policies, it has generally attempted to drive borrowing countries towards commercialization and privatization of their energy sectors *before* any sound basis for open discussion of energy policy is in place. In practice, this has meant that privatization has stifled the possibilities for open discussion of energy policy in most countries. This is partly because the concept of an openly discussed 'national energy strategy', designed to meet the public interest, is overtaken by the concept that 'leaving it to the market' will meet the public interest. In addition, open discussion is hindered because in competitive energy markets information on private deals is not usually made public (let alone subjected to open public debate), even after such deals have been finalized.

The Bank is not ignorant of its own failure to ensure effective regulation. The internal Bank review of energy lending, mentioned above, which was produced in February 1988, stated clearly that:

> Progress is ... slow on establishing sound regulatory frameworks for the power subsector in many countries, and some governments have attempted to alter the rules of the game after the private sector has entered the market

... Many countries have regulation as good as those in developed countries, but compliance and enforcement are poor ... Borrower commitment to compliance and enforcement of regulations has been and continues to be a major impediment to significantly reducing pollution levels.[18]

No Shift Towards Sustainable Energy

In recent times, the Bank has become quite candid in its *internal* documents in admitting its lack of attention to sustainable energy. In relation to the environment, the Bank's February 1998 internal review of its energy lending, for example, states that:

In theory, opening the market to private producers should allow for small-scale electricity production (small hydro and other renewable energy) and cogeneration ... Experience to date is [however] inconclusive ... further research is required to determine whether the new regulatory frameworks being developed are making a level playing field for cogeneration[19] and renewable energy to compete ... [Meanwhile, the available evidence shows that] ... to date, there has been little power generation from renewables or geothermal where the power sub sector has been restructured.

The facts are that, despite its reform programme, which focuses on privatization of the energy sector, the Bank has paid little attention to sustainable energy. A total of $429 million was lent by the Bank for six renewable energy projects from 1980 to 1993 and $172 million for four projects from 1994 to 1997. This is only 1.4 per cent of total energy sector lending. The Bank's 1998–2000 pipeline includes 14 projects that contain renewable energy components; total lending for these projects is $620 million. This is only 3.6 per cent of total energy sector lending. The Bank's experience with the development of wind power is limited to India, where it has installed 13 schemes with a total capacity of only 41 MW. Further, Bank lending for 'non-traditional renewables' such as solar home systems (solar panels) has been small in the past, and even in the last four years the Bank has invested in only four projects. This represents just 1.4 per cent of total energy sector lending. With the increased emphasis on privatization in the Bank's energy policies, the International Finance Corporation – the Bank's private lending arm – became increasingly involved in lending for the power sector during the 1990s. However, only about 7 per cent of the IFC's energy lending was for renewables during that period and the IFC has made few investments in energy efficiency-related projects. The IFC has concentrated mainly on investments for the exploration and development of oil and gas, and diesel and gas generation projects.[20]

The Bank has contributed to a general bias towards fossil fuel lending (for oil, gas and coal) because of the way it has selectively implemented its policies, concentrating mainly on privatization, without concurrently ensuring adequate regulation. The Bank in fact conceded this in a virtual consultation on energy and environment, which it held with NGOs, academics, donors and others on the Internet during 1997. The Bank admitted that:

> Despite expected positive long-term effects of deregulation, unregulated electricity markets are likely to put renewable energy technologies at a disadvantage in the short-run because they favor the cheapest energy as determined purely by price, but do not capture environmental and social externalities. In addition, investors in unregulated, short-term markets favor low capital-cost fossil fuel technologies over renewables that tend to be more capital-intensive, even if they may produce more cost-effective power over their lifetime.[21]

New and renewable energy technologies are still perceived by most investors as having large commercial risks, so they are relatively expensive to finance. Also, energy efficiency and renewable technologies are disadvantaged by the massive subsidies with which governments support the fossil fuel and nuclear industries. So, for many reasons, the way the Bank is implementing its policies for reform of the energy sector is leading to more favourable conditions for conventional fossil fuel technologies, which offer the most profit in the short term, while placing renewable energy and energy efficiency technologies at a disadvantage. Moreover, as mentioned above, most energy development is taking place for urban industrialization, while extension of energy services into rural areas is largely being ignored.

Since the Bank's 1992 energy policies were adopted, a succession of independent reports have criticized the Bank for not meeting its promise to provide sustainable energy. These reports are described in more detail below. It is important to stress here, however, that the Bank itself has now begun to admit, at least in reports produced for internal discussion, that there has been no substantial shift in Bank investments towards sustainable energy.

So where, then, has the Bank been putting the bulk of its investments? In most regions, since the Bank's 1992 energy policies were adopted, total energy sector lending has remained stable or has even declined. This is a result of the push for privatization. As the private sector has been expected to increase its investments in power generation (such as power plants), so the Bank has focused more on relatively less expensive activities, such as giving technical advice to governments on how to privatize, and on lending for energy transmission and distribution. The main exception for the Bank has been its investments in Russia, where it has joined a frenzy of lending for oil and coal exploration, mining and generation, and where, in the four years after its 1992 energy policies were adopted, its investments doubled compared to the previous four years.

The slowdown in Bank energy investments has also been due to citizens' campaigns against the negative social and ecological impacts of large hydroelectric dams. Many such campaigns have succeeded in stalling or suspending large dam projects. The debacle of the Narmada Sagar and Sardar Sarovar dams in India, which were both eventually suspended, for example, brought much negative publicity for the Bank. Other uncompleted projects include the Rogun dam in Tadijikstan, the Machadinho and the Porto Primavera dams in Brazil, or the Nam Choan dam in Thailand.[22] Publicity surrounding such projects has exposed massive environmental damage and widespread human rights abuses, with the consequence that large dam projects are now perceived as high-risk investments and lending for dams has decreased.

Total World Bank lending for the energy sector between 1980 and 1997 is shown in Table 2.1. Bank lending for coal has shifted, in line with its 1992 policies, away from coal exploration and coal mine development

TABLE 2.1 World Bank lending for the energy sector, 1980–97[23]

Year	Power	Oil and gas	Mining and other	Total energy	Total bank lending	Lending %
1980	2,650.3	358.0	179.5	5,167.8	11,441.7	28
1981	1,656.0	649.5	22.9	4,218.4	12,210.6	18
1982	2,181.2	945.7	260.5	5,369.4	13,017.9	26
1983	1,913.2	820.0	336.8	5,053.0	14,388.6	21
1984	2,692.8	630.7	240.4	5,547.9	15,320.4	23
1985	2,432.3	667.4	506.0	5,590.7	14,233.7	35
1986	2,811.9	249.1	3.9	5,050.9	16,565.7	19
1987	3,037.9	327.4	347.0	5,699.3	18,085.2	21
1988	2,048.4	358.0	79.5	4,473.9	19,156.7	13
1989	3,064.1	799.5	50.0	5,902.6	21,299.4	18
1990	2,968.3	336.0	25.0	3,329.3	20,702.0	16
1991	1,706.8	1367.4	221.0	3,295.2	22,686.0	15
1992	3,554.3	518.6	6.0	4,078.9	21,708.0	19
1993	2,488.1	849.2	262.0	3,599.3	23,696.0	15
1994	1,613.3	1143.5	14.0	2,770.8	20,836.0	13
1995	2,241.5	603.1	24.8	2,869.4	22,522.0	13
1996	3,247.1	55.6	692.0	3,994.7	21,517.0	19
1997	1,889.2	6.4	450.6	2,346.2	19,146.0	12

Source: *The World Bank Environment Strategy for the Energy Sector: an OED Perspective*, World Bank, 1998.

towards technical assistance for privatization of the coal industry, particularly in Central and Eastern Europe and in Russia. Further, at the time of writing the Bank is gearing up for a major increase in lending for oil and gas. Bank lending in the period 1998–2000 for oil and gas is expected to be 14 per cent higher than the annual average for the period 1990–97.

As already explained, the World Bank's 1992 energy policies focus on the encouragement of private sector involvement in the energy sector. This has not gone smoothly, as described earlier. Nevertheless, the Bank's efforts have led to a substantial increase in private participation, and it is important to understand where and on what this money is being spent.

Between 1990 and 1997 private sector involvement in the energy sector grew substantially. Private companies undertook the management, operations, rehabilitation, or construction risk of 534 projects, with investments totalling US$131 billion. Most of these projects were concentrated in East Asia, where 165 private projects represented a total investment of US$ 50 billion, and in Latin America and the Caribbean, where 169 private power projects represented a total investment of US$45 billion.[24]

Africa, which is arguably the region suffering the most from a lack of available energy, has been largely left behind by private investors, who have tended to see the continent as high risk. Private investment in the CEE region, meanwhile, has tended to concentrate on improving the supply-side efficiency and reliability of existing power plants, rather than on building new generation capacity.

Under the influence of the Bank's 1992 energy policies, privatization

TABLE 2.2 Top ten countries attracting private participation in electricity, 1990–97[25]

Country	Total investment in projects with private sector participation (US$ millions)	Projects
Brazil	17,644	20
China	15,015	60
Argentina	12,011	63
Philippines	10,901	39
Indonesia	9,569	13
India	9,219	29
Pakistan	6,924	20
Malaysia	6,330	9
Colombia	5,873	16
Thailand	5,645	39
TOTAL	99,130	308

Source: Private Participation in Infrastructure (PPI) Database, World Bank.

has been spreading among all low-income countries. Nevertheless, it is striking that most of the investments have been captured by only a small number of countries. In fact the 'top ten' low-income countries accounted for more than three-quarters of all private energy investments between 1990 and 1997 (see Table 2.2).

Another disturbing fact is that almost three-quarters of private investment in energy in low income countries between 1990 and 1997 has gone into constructing new power generation plants using fossil fuels (oil, coal or gas), while the remaining 25 per cent has gone into existing energy utilities, or transmission and distribution projects. More than half of all private investment in energy in low-income countries between 1990 and 1997, moreover, has gone into 286 new power generation projects on previously undisturbed, so-called 'greenfield', sites.[26] Meanwhile, only a relatively tiny amount of private investment has been made in energy efficiency or renewable energy.

In short, the way the World Bank has implemented its 1992 energy policies has induced the private sector to participate in the energy sector mainly in the 'top ten' low-income countries in Asia and Latin America, where private investments have created a proliferation of new fossil fuel power plants, mainly on greenfield sites.

Meanwhile, most of the world's low-income countries are still suffering from financial crises in their energy sectors, new and renewable energy technologies are being marginalized, and the world's two billion rural poor – particularly those in sub-Saharan Africa, which is attracting the least private sector investments, but also those in most other low-income countries – have gained little or no benefits.

The Bank versus Civil Society

The Bank's 1992 energy policies were adopted in the wake of the 1992 Earth Summit, where leaders from almost every country in the world had pledged their support for the environmental cause. There was a certain amount of cautious optimism at that time among NGOs that, perhaps, a real change in the direction of sustainability might be finally taking place. The Bank's 1992 energy policies contained many promises of sustainable energy, expectations had been built up by the Bank, and NGOs expected to see substantial changes in the Bank's operations in the direction of sustainable energy. As described above, however, the Bank continued largely with business as usual, and this soon became clear to NGOs.

The first independent report to make the point, *Power Failure*, was produced in March 1994 by two Washington, DC-based NGOs, the Environmental Defense Fund and the Natural Resources Defense Council.[27]

Both organizations had publicly welcomed the Bank's policies when they were produced, mainly because the policies stress the importance of integrated resource planning (see Box 2.3). *Power Failure* focused on the four most important requirements contained in the Bank's policies. These were for electric power loans:

- to be based on integrated energy strategies;
- to contain significant components to build the institutional capability of low-income countries to address energy efficiency;
- to require movement toward economic pricing; and
- to establish or strengthen government regulatory systems that address improved end-use efficiency.

Power Failure reviewed all Bank power loans under preparation during the first six months of 1993, and then graded them for compliance in the four areas mentioned above. The conclusion was that out of the 46 loans under preparation for 33 countries, only two loans were set to comply with the Bank's energy policy, while only three others contained comprehensive support for improved end-use energy efficiency. The report found that Bank staff did not have any requirement or incentive to operationalize the policies, so they were applying it selectively, giving most attention to privatization and pricing. The key findings were as follows:

Finding 1: There is no incentive for, or requirement of, Bank task managers to give end-use energy efficiency a high priority in power loans ...

Finding 2: Few loans address end-use energy efficiency through any means other than price increases ...

Finding 3: There are only a few cases in which World Bank loans are attempting to incorporate demand side management. ..

Finding 4: World Bank staff sometimes misunderstand integrated resource planning, or underestimate its applicability to developing countries ... Common misconceptions include ... a) IRP is a uniquely US approach, and thus, inapplicable in other countries ... b) IRP is too sophisticated for developing countries ... c) IRP involves excessive government intrusion in free markets ... d) Developing country governments and utilities will resist IRP (and thus should not be required to do it) ... e) IRP/DSM requires electricity ratepayers who do not participate in energy conservation programs to pay for those who do ...

Finding 5: There is a perception by many Bank power staff, that other power sector reforms, such as pricing and privatization, deserved higher priority than do measures designed explicitly to improve end-use energy efficiency ...

Finding 6: A new unit [the Asian Alternative Energy Unit, ASTAE], established at the Bank, to help incorporate end-use energy efficiency in Asian energy loans, has shown promise, but has been hampered in much of its efforts ... bureacratic conflicts between ASTAE, the region's country departments, and the Bank's central Industry and Energy Department appear to have stymied potential cooperation.

The report ended with recommendations designed to help the Bank address its problems in promoting sustainable energy. In brief, these were: to make the energy policies binding on all staff; to increase investment in end-use efficiency; to consider end-use efficiency on an equal footing with other power sector reforms (such as pricing and privatization); to include end-use efficiency in environmental assessments; to increase the internal capacity to address IRP, DSM and end-use efficiency tools and techniques; to develop a climate change policy; and to conduct independent reviews of Bank progress in end-use efficiency.

Power Failure's conclusions have, of course, been proved largely correct by more recently produced internal World Bank reports, as shown above.[28] The Bank's reaction to the report's criticism and suggestions, however, was hostile and defensive in the extreme. In the press, the Bank's Industry and Energy Department viciously attacked the authors and attempted to discredit the report. A statement released by the Bank on 21 March 1994, for example, declared that:

> The World Bank rejects the conclusions of the report in their entirety. It deliberately falsifies the positions of the World Bank, whose lending and policies are very much concerned with the efficient provision and use of energy ... [the authors] fail to understand the situation of billions of people in the world and what is needed for energy efficiency and development ... [the authors] are totally insensitive to the needs of developing countries ... Two billion people in developing countries have no access to electricity whatsoever. How many of these people can use, let alone afford, the \$25 light bulbs or the high efficiency air conditioners that rich country lobby groups say are so essential for energy efficiency? ... These rich country lobby groups advocate policies that would deny the world's poor access to modern fuels ... The World Bank will therefore continue its efforts on both the demand and supply sides to help improve the energy situation in the poorest countries of the world, on which the prosperity of these countries depends.[29]

Thus the Bank damned *Power Failure*. The report had raised the alarm, nevertheless, among citizens' groups around the world. In Sweden, for example, a network of 30 NGOs active in development and environment issues had already formed the Världsbanksforum (World Bank Forum), within which energy had become a key issue of interest. In 1995 one of the forum's members, the Swedish office of the Worldwide Fund for Nature (WWF), commissioned an independent Swedish consultancy, JGT Associates, based in Stockholm, to produce a report on the Bank's implementation of its energy policies. Research for the report, *A Negawatt Saved*, was carried out during May 1995 and the report was produced in February 1996. The term 'negawatts' refers to the vast possibilities for energy savings, or negative watts. The report took a sample of 56 World Bank energy projects, which were in various stages of preparation or recently approved; it then analysed the extent to which these projects adhered to the requirement in the Bank's 1992 energy policies that: 'lending in the energy sector should be based on and support the development of integrated energy strategies that take advantage of all energy supply options, including cost-effective conservation-based supplies and renewable energy sources'.[30]

Since the Bank had still not shifted its investments in the direction of sustainable energy in 1995, it is not surprising that *A Negawatt Saved* reached the same general conclusions as *Power Failure*. It found that of the 56 projects which were sampled, only three fully complied with the Bank's energy policy. *A Negawatt Saved* further noted that:

> Generally, the World Bank's efforts in pursuing integrated energy strategies have been inconsistent. Only a small fraction of project investment planning considers demand-side resources or renewables, while in the background almost all of the potential savings are disregarded in the Bank's standard investment planning for the power sector. The status quo of supply-side, fossil fuel-fired energy development persists – even though there is a gold mine of 'negawatts' available, without any of the social and environmental externalities that plague the big power generation projects.
>
> World Bank and upper management and Executive Directors face the choice of whether or not to help position the Bank as a global leader in the transition to sustainable energy development. The Bank will inevitably be a leader in one way or another, either for progress or for business as usual. Despite tentative progressive efforts by a few divisions and individuals within the Bank, operations still overwhelmingly support the status quo.

It was clear to NGOs by now that the Bank's 1992 energy policies were being implemented selectively. Generally speaking, the Bank's staff were carrying out the requirements relating to pricing reforms and privatization while ignoring the requirements related to integrated resource planning and energy efficiency.

The problem of staff selectively implementing policy had, in fact, long been known within the Bank. Indeed, an attempt was already under way to deal with the problem through a series of internal institutional reforms. Bank staff had previously been guided by internal policy documents called *Operational Directives* (*OD*), and, earlier, *Operations Manual Statements*. A commitment to streamline the older documents was made, however, as a result of an independent report commissioned by the Bank, called *Effective Implementation: Key to Development Impact*, which was produced in September 1992. The commissioning of this report was largely triggered by widespread public concern and criticism of the Bank's investments in the Sardar Sarovar and Narmada Sagar hydrolectric dam projects in India, both of which carried appalling ecological and social impacts, and from both of which the Bank eventually withdrew. The report was prepared by a specially set up Task Force on Portfolio Management, chaired by Willi Wapenhans, a former vice-president of the Bank, and became known as the Wapenhans Report. It was highly critical of various aspects of the Bank's operations and pointed out, among many other things, that *it is almost impossible to hold Bank staff accountable for following Bank policies*. The Wapenhans Report concluded that the older *Operational Directives* and *Operational Manual Statements*, mentioned above, were so long and complicated that Bank staff could not tell the difference between the 'bottom line' of mandatory policy and 'wouldn't it be nice to have' statements of intentions. As a result of the Wapenhans Report it was decided to move to a new format of *Good Practices* (*GP*), *Bank Procedures* (*BP*), and *Operational Policy* (*OP*). Each of these new documents was intended to be concise, to the point, and clear on what Bank staff should and should not do in their activities for the Bank. The status of these new documents was to be simple and straightforward: *Operational Policy* and *Bank Procedures* would be mandatory, while *Good Practices* would be non-mandatory guidelines only.

During 1995 all of the Bank's departments were required to put their policies through a process of 'conversion' from the old *OD*s to the new *OP*s, *BP*s, and *GP*s. The Bank's 1992 energy policies only got as far as an internal draft for comments (Draft OP4.45), however, before becoming stalled in internal disagreements between different groups and individuals within the Bank. By the spring of 1996, it became evident from a leaked internal Bank memorandum that the conversion process had highlighted 'profound and enduring differences' between different parts of the Bank and that some Bank staff were even complaining that 'we have to be increasingly careful in setting policy that we are able to implement in practice'.[31]

In the midst of this internal wrangling, early in the summer of 1996, the Bank finally did convert its energy policies. Not into mandatory *Operational Policy*, however (as it was required to do by its own Operations Evaluation Department), but rather into non-mandatory *Good Practice* documents (known as *GP4.45* and *GP 4.46*).[32] This move effectively down-graded the importance of the 1992 energy policies. In one stroke the energy policies became entirely non-binding. Moreover, crucial elements of the policies, relating to sustainable energy, were not included in the new *Good Practice* documents at all. Evidently, the vested interests supporting business as usual within the Bank had succeeded. There would be no mandatory requirements for staff to implement the policy requirements related to sustainable energy.

Given the institutional history since the Wapenhans Report, described above, and given the Bank's already poor record on sustainable energy, the Bank's move to hold its staff even *less* accountable for implementing its energy policies caused huge public concern and disappointment – so much so that in August 1996 28 NGOs from North America, Latin America, Western Europe, Central and Eastern Europe, Africa and Asia sent a jointly written, open letter to the Bank's president, James Wolfensohn, in which they stated:

> We are writing to express in the strongest possible terms our deep concern and disappointment concerning the recent issuing of new Bank energy policies as non-binding 'Good Practice' documents (GP4.45 'Electric Power Sector; GP 4.46 'Energy Efficiency') rather than as binding Operational Policies and Bank Procedures. We are also deeply concerned that the conversion process has removed several crucial elements of the original Board approved policy papers (The World Bank's Role in the Electric Power Sector and Energy Efficiency and Conservation in the Developing World). We believe that the Bank's actions reflect a lack of commitment to sustainable energy – in stark contrast to its stated goals … we would like an explanation of the Bank's 'eleventh hour' downgrading of the 1992 Board-approved policy papers to 'Good Practice' documents. Most importantly, we want a commitment from you that the Bank keep its promise of reissuing the 1992 Board-approved policies as binding Operational Policies that accurately represent the content of the original documents … Effective implementation of the Bank's energy policies now is crucial. When the policies themselves are non-binding and hortatory, the Bank's commitment to effective implementation of them is questionable.

Unfortunately, no commitment from James Wolfensohn was forthcoming. The Bank's reply came, rather, in September 1996, from Jean François Rischard, the vice-president of the Finance and Private Sector Development. This is revealing in itself: while other departments, such as Agriculture, Water, and Transport, for example, were located within the Presidency for Environmentally Sustainable Development, the Bank's Industry and Energy Department was located within the presidency for

Finance and Private Sector Development, which is the Presidency chiefly responsible for privatization. In a soothing tone, Jean François Rischard, one of the Bank's most senior staff responsible for privatization, explained to NGOs that:

> we believe that your concerns … stem largely from a misunderstanding about the process of preparing and issuing such instructions … there has been no downgrading of the Bank's energy policies. The Bank policies on electric power and energy efficiency continue to be those detailed in the two documents approved by the Board in 1992 (the Power Paper and the Efficiency Paper). These two Bank Policy Papers state the Bank's policies: they were circulated to staff and are widely used and referred to in our energy operations … we remain committed to follow a consultative process in implementing our energy policies. We intend to continue our outreach efforts in that respect.

Less than one year after Rischard's letter, however, the centrepiece of the Bank's efforts to reach out and dialogue with its critics confirmed the U-turn on the status of its 1992 policies. The *Preliminary Draft Energy and Environment Strategy Paper* was part of a dialogue with 'external stakeholders', which the Bank began in 1996, immediately after its exchange with the global group of NGOs mentioned above. The *Strategy Paper* has since evolved into the Bank's official *Energy and Environment Strategy*, known as *Fuel for Thought*. The draft version of *Fuel for Thought*, just like the 1992 energy policies, contained a number of promising recommendations for sustainable energy. But the Bank left no room this time for 'misunderstanding'. It blatantly declared that the *Strategy Paper* had been written because the *Power Paper* and the *Efficiency Paper* (the Bank's board-approved 1992 energy policies), were in fact only 'formal papers' which 'did not, nor were meant to, set out a strategy to address energy-environment issues directly'.[33] This statement is clearly part of the Bank's U-turn on its 1992 energy policies. To begin with, it is clearly misleading to suggest that the Bank's energy policies are only 'formal papers'. Secondly, there is no question that the energy policies did indeed lay out a direct strategy to deal with energy–environment issues. The *Power Paper*, for example, states that:

> to assist in mobilizing resources, the Bank must help (client) countries put in place systems that encourage the efficient use of power-sector resources. This efficiency objective also helps developing country power sectors meet their local and global environmental responsibilities as they become larger components of the world energy market.

The *Efficiency Paper* begins its Executive Summary, moreover, by stating that there is 'a congruence of several forces in the developing world that makes very timely the formulation of a strategy to address energy efficiency and conservation issues better'.

The *Efficiency Paper* then goes on to outline such a strategy. Chapter 5 deals with country policy priorities, Chapter 6 with sector priorities. Chapter 7 is even called 'Strategy for the World Bank', and goes on to state in detail how the Bank will better integrate energy efficiency issues into its country policy dialogues, will be more selective in its lending, and will identify, support, and give high-level in-country visibility to, approaches for addressing demand-side management.

The Bank's new environmental strategy for the energy sector, *Fuel for Thought*, which is described in more detail below, is therefore a compromise born out of internal disagreements within the Bank and part of a successful attempt by certain staff to downgrade the Bank's 1992 energy policies *in order to prevent any mandatory requirements to implement the integrated resource planning and energy efficiency principles contained in the Bank's 1992 energy policies.*

In May 1997, pressure on the Bank was maintained when another independent report was produced, which backed up the findings of both *Power Failure* and *A Negawatt Saved*. The study, *Aiding Global Warming: an Analysis of Official Development Assistance for the Coal Industry*, was produced by the Australian NGO AidWatch. The scope of the study included not only the World Bank but also the Asian Development Bank, the Global Environment Facility (GEF, mentioned above) and the Australian government's overseas development agency, AusAID. In relation to the World Bank the report found that:

> The World Bank continues to focus on large-scale, supply-side energy developments, including significant investments in expansion of the grid in low income countries. Over the past two years the World Bank has invested $2.24 billion in financing of new coal fired power stations, rehabilitation of existing coal fired power stations and district heating centres, and restructuring of coal mining industries. It is significant that in terms of lending for new capacity additions, coal continues to attract the largest proportion of World Bank funding.
>
> In comparison, over the past two years the World Bank spent only $61 million on demand side management (DSM), and a mere $5.9 million on one renewable project. Thus the Bank spent thirty three times more money on coal than on renewables and DSM over the last two years. The Bank claims that it is currently increasing its funding of renewables and DSM. However, an analysis of the Bank's projects in the pipeline for 1997 and beyond reveals that the Bank will still be lending around $1.5 billion more

for coal projects than for demand side management and renewable energy projects in the next 2–3 years.

Similarly, the Global Environment Facility [which is based within the World Bank] grant portfolio, which operates the financial mechanism of the Framework Convention on Climate Change on an interim basis, shows a bias towards funding for fossil fuels. In the period FY1995–FY1996, the GEF approved grants of $34.2 million for fossil fuels as opposed to $18.4 million for projects involving renewable energy technologies and/or DSM. This is of particular concern given that one of the GEF's major priorities is supposedly reducing greenhouse gas emissions.

AidWatch's report, *Aiding Global Warming*, recognized that fossil fuels (coal, oil and gas) are a major energy source for many low-income countries.

Box 2.4 The Climate Convention and Kyoto Protocol

In response to the threat of climate change, international negotiations have produced the United Nations Framework Convention on Climate Change (UNFCCC), which came into force on 21 March 1994. The UNFCCC's main aim is to reach 'stabilization of greenhouse gas concentrations in the atmosphere at a level that would prevent dangerous anthropogenic interference with the climate system'. This level should be reached fast enough to allow ecosystems to adapt in a natural way to climate change, to ensure that food production is not threatened, and to ensure that economic development is able to proceed in a 'sustainable' manner. The convention rests on the principle of 'common but differentiated responsibilities', which recognizes that all nations are responsible for protecting the climate, but that the industrialized countries are chiefly responsible for most of the greenhouse gas build-up that has occurred to date. The low-income countries have been least responsible for the present threat of climate change, yet it is these countries that are most vulnerable and threatened by the effects. Moreover, the low-income countries are least able to pay for greenhouse gas mitigation. This principle takes shape in a commitment that Annex II (donor) countries should provide new and additional financial resources through the Global Environment Facility (GEF), the Convention's 'financial mechanism', and should carry out socially and economically beneficial technology transfer. Further, the convention makes it clear that energy consumption in the low-income countries must grow in order to achieve 'sustained eco-

The report nevertheless recommended that development money be used to highlight the significant benefits that can be achieved by moving away from fossil fuels. Specifically, the report recommended that the AusAID programme should be dedicated to energy efficiency and renewables, with special attention to funding indigenous renewable energy initiatives and the capacity-building of local communities in low-income countries; that the World Bank and Asian Development Bank should place an immediate moratorium on lending for new coal-fired power stations and large dams and should vigorously promote demand-side management and small-scale decentralized renewable energy technologies over large-scale supply-side projects; that all energy sector lending should be in the context of integrated resource planning (as stipulated in the Bank's 1992 energy policies); and

---

nomic growth and the eradication of poverty'. Meanwhile Annex I parties (OECD countries plus the 'economies in transition') should aim for stabilization of their greenhouse gas (GHG) emissions. Low-income countries have no mandatory GHG emission restrictions; industrialized countries are supposed to limit their GHG emissions.

On 11 December 1997, the Kyoto Protocol to the UNFCCC was adopted. If it enters into force it will mean binding CO_2 reduction limitations for 39 industrialized countries and countries with 'economies in transition' (Annex B countries under the protocol). These countries agreed to make sure that their aggregate GHG emissions do not exceed their individually assigned amounts, 'with a view to reducing their overall emissions of such gases by at least 5.2 per cent below 1990 levels in the commitment period 2008–12'. The Kyoto Protocol allows some flexibility for Annex B countries to find ways to meet their obligation; this has become known as the 'flexible mechanisms'. The countries are allowed to trade carbon reductions among themselves (so called emissions trading), for example. They can also jointly implement projects that can lead to carbon reductions on a project basis by reducing emissions or improving sinks (so called joint implementation or JI). Emissions trading is allowed only among Annex B countries. The same goes for joint implementation, except that JI can also take place with low-income countries (non-Annex B countries) under the clean development mechanism (CDM). Starting from the year 2000, 'carbon crediting' is allowed for JI projects.

that the GEF should focus solely on renewable energy and demand-side management.

Hard on the heels of the AidWatch report came *The World Bank and the G7: Changing the Earth's Climate for Business*, which was produced collaboratively by the Institute for Policy Studies, USA, the International Trade Information Service, USA, the Halifax Initiative, Canada, and the Reform the World Bank Campaign, Italy.[34] This report analysed World Bank fossil fuel lending since the 1992 Earth Summit. It was first published in June 1997 and has since been followed by a number of updates. The latest version at the time of writing contains the following key findings:

1. Since the 1992 Earth Summit, the World Bank has spent 25 times more on climate-changing fossil fuels than on renewables. The fossil fuel projects the World Bank has financed will over the next 20 to 50 years add carbon dioxide emission to the Earth's atmosphere equivalent to 1.3 times the total amount emitted by all the world's countries in 1995.

2. World Bank President James Wolfensohn's 1997 Earth Summit II pledge to calculate greenhouse gas emissions associated with World Bank projects has proven hollow: Less than 10 percent of all World Bank projects are being calculated for their impact on the climate.

3. Since the 1992 Earth Summit, nearly one in five World Bank fossil fuel dollars promoted coal and diesel-fired power plants in China; from 1997–98, one in three did.

4. The World Bank is becoming dangerously entangled in a web of conflicting economic and political interests, the outcome of which may play a deleterious role in the stability of the Earth's climate while making the poorest worse off.

5. Although earmarked for sustainable development and poverty relief, 9 out of 10 World Bank fossil fuel projects benefit transnational corporations based in the wealthy G-7 countries, many of whom are members of a U.S. based lobbying group, the Global Climate Coalition, that actively opposes any action on climate change.

6. The World Bank is helping open up some of the world's richest untapped oil and gas fields in regions ruled by dictators, while ignoring renewable energy opportunities that others are seeking out.

As mentioned earlier, the Bank's main effort to stave off criticism of its lack of investments in sustainable energy has been its *Energy and Environment Strategy Paper*, also known as *Fuel for Thought*.[35] This paper provides an analysis of energy, environment and development and lays out an action plan for the Bank. The *Strategy Paper* contains many promising suggestions, but it is weakened by its lack of attention to the internal institutional constraints within the Bank, in particular the lack of support for sustainable energy issues among Bank staff. The Bank's original plan was to finalize and present *Fuel for Thought* for Board approval during 1997. Because of disagreement between different groups and individuals within the Bank, however, which slowed the process, *Fuel for Thought* was not presented to the Bank's board of directors until November 1998. By that time it had become so watered down that the Bank's board rejected it for failing to lay out any practical steps that the Bank should take in the direction of sustainable energy. The Board finally approved the paper on 20 July 1999, despite criticisms from, for example, the US government representatives to the Bank, who issued a formal statement noting that: 'We have the impression that the exercise did not really change the Bank's way of doing business in the energy sector'[36] (see Chapter 18).

These criticisms are mainly focused on the section in *Fuel for Thought* called 'Monitorable Progress Indicators'. The outcomes listed within this section – which was inserted only under pressure from the board – underline the Bank's vision of energy sector privatization as an end in itself, rather than as a means to promote sustainable energy. Listed 'outcomes' are: wider more competitive energy markets, regional energy trade, and private participation in energy utilities. Amazingly, oil exploration is also listed as a monitorable outcome within the new strategy.

Further, the Bank lists the completion of a number of already approved renewable energy and energy efficiency projects as monitorable outcomes, in addition to a small number of sustainable energy projects that were destined to take place anyway, and a number of other projects that were designed to ameliorate some of the environmental damage caused by mining.

In fairness, a number of new projects for energy efficiency and renewable energy are listed as outcomes in *Fuel for Thought*. But these remain small in relation to the Bank's total portfolio. And, taken as a whole, the monitorable progress indicators in *Fuel for Thought* does not represent any substantial shift away from what the Bank has been doing throughout the 1990s: concentrating on the privatization of the energy sector as an end in itself, focusing mainly on lending for coal, oil and gas, and implementing a relatively small number of promising sustainable energy initiatives on the periphery, which mainly have come about because of political and financial support from outside the Bank.

Further, on the crucial issue of holding Bank staff accountable for following Bank policies, *Fuel for Thought* is almost silent. The paper mentions briefly that mandatory Operational Policies for the Energy Sector will be issued 'in due course'. But no dates are set, and the creation of mandatory *Operational Policy* for the energy sector is absent from *Fuel for Thought*'s list of monitorable progress indicators.

This means that neither the Bank's 1992 energy policies nor *Fuel for Thought* itself are mandatory policy within the Bank, and the selective implementation of policy by Bank staff – favouring privatization as an end in itself rather than any direct measures to promote sustainable energy – is continuing. The way in which this has translated into specific investments in the energy sectors in low-income countries is described in the country studies that make up the following section of this book.

Notes

1. Glen, J. D. (1990) *Private Sector Electricity in Developing Countries: Supply and Demand*, Discussion Paper no. 15, International Finance Corporation, Washington, DC: World Bank.

2. Martin, B. (1993) *In the Public Interest? Privatization and Public Sector Reform*, London: Zed Books.

3. World Bank (1993) *The World Bank's Role in the Electric Power Sector*.

4. Ibid.; and World Bank (1993) *Energy Efficiency and Conservation in the Developing World*.

5. World Bank (1993) *Energy Efficiency and Conservation in the Developing World*.

6. World Bank (1998) *The World Bank Environment Strategy for the Energy Sector: an OED Perspective*.

7. Ibid.

8. See, for example, Bruggink, J. (1997) 'Market metaphors and electricity restructuring', *Pacific and Asian Journal of Energy*, TERI, vol. 7, no. 2, December.

9. World Bank (1999) *Project Appraisal Report on Andhra Pradesh Power Sector Restructuring Project*.

10. From the Large Public Power Council website, quoted in Heidenreich, A. (1998) *Public versus Private Power in the North Country*, NGONet/Third World Institute, http://www.lppc.org.

11. R. Munson, *The Power Makers: The Inside Story of America's Biggest Business … and its Struggle to Control Tomorrow's Electricity*, Emmaus, PA: Rodale Press, 1985.

12. Heidenreich, A. (1998) *Public versus Private Power in the North Country*, NGO Net/Third World Instititute.

13. Bruggink, J. (1997) ' Market metaphors and electricity restructuring', *Pacific and Asian Journal of Energy*, TERI, vol. 7, no. 2, December.

14. World Bank (1998) *The World Bank Environment Strategy for the Energy Sector: an OED Perspective*.

15. World Bank (1998) *The World Bank Environment Strategy for the Energy Sector: an OED Perspective*, for example, quotes from a review of the Electric Power Research Institute and the 1992 *Annual Review of Energy and the Environment* to assert that potential savings from DSM have been considerably overestimated.

16. *National Report for India, MDB–Energy Project*, Both ENDS, 1999.

17. World Bank (1996) *Rural Energy and Development: Improving Energy Supplies for Two Billion People*.

18. World Bank (1998) *The World Bank Environment Strategy for the Energy Sector: an OED Perspective.*

19. Co-generation is the combined production of heat and power.

20. World Bank (1998) *The World Bank Environment Strategy for the Energy Sector: an OED Perspective.*

21. World Bank responses to comments received via the virtual consultation on the energy and environment strategy discussion paper of 22 July 1997, section 33.

22. McCully, P. (1996) *Silenced Rivers: The Ecology and Politics of Large Dams*, London: Zed Books.

23 World Bank (1998) *The World Bank Environment Strategy for the Energy Sector: an OED Perspective.*

24. Izaguirre, A. K. (1998) 'Private participation in the electricity sector – recent trends', in *Public Policy for the Private Sector*, World Bank, December..

25. Private Participation in Infrastructure (PPI) Database, World Bank.

26. Izaguirre, A. K. (1998) 'Private participation in the electricity sector – recent trends', in *Public Policy for the Private Sector*, World Bank, December.

27. Environmental Defense Fund and Natural Resources Defense Council, *Power Failure: a Review of the World Bank's Implementation of its New Energy Policy.*

28. Such as: World Bank (1998) *The World Bank Environment Strategy for the Energy Sector: an OED Perspective.*

29. World Bank (1994) World Bank, *Statement in Response to EDF/NRDC Study on World Bank Energy Policy*, 21 March.

30. World Bank Policy Paper (1993) *Energy Efficiency and Conservation in the Developing World.*

31. World Bank (1996) Office Memorandum on *Conversion of Remaining Operational Directives*, World Bank, 15 March.

32. GP4.45 Electric Power Sector, GP4.46 Energy Efficiency.

33. *Energy and Environment Strategy Paper*, Provisional Draft, 22 July 1997.

34. Sustainable Energy and Economy Network (Institute for Policy Studies, USA), International Trade Information Service (USA), Halifax Initiative (Canada), Reform the World Bank Campaign (Italy), *The World Bank and the G-7: Charging the Earth's Climate for Business: An Analysis of the World Bank Fossil Fuel Project Lending since the 1992 Earth Summit*, June 1997.

35. World Bank (1999) *Fuel for Thought.*

36. US government statement on *Fuel for Thought*, 1999.

Part II

World Bank Energy Policy in Practice: Country Studies

This section describes the energy situation and impacts of World Bank and other MDB energy policies and investments in the African, Asian, Central and Eastern European, and Latin American and Caribbean regions. From Brazil to Zimbabwe, these country studies reflect the concerns of NGOs that have first-hand experience of the activities of the World Bank and other MDBs in the energy sector in their countries.

3

Cameroon[1]

Energy and Economic Crisis

Since the beginning of the 1980s Cameroon has been in the grip of an economic crisis and the already low purchasing power of the population was aggravated by the devaluation of the currency in January 1994. Given this situation, Cameroon is dependent upon intervention by government and international development agencies, such as the World Bank, to open new investment in sectors that have been largely ignored since the beginning of the country's economic crisis, such as education, health and agriculture. Investment is needed to deal with the whole range of social and environmental problems facing the country, including problems related to energy.

Cameroon has significant reserves of oil and gas. Coal and uranium are present, but in smaller quantities. Potential renewable energy resources are available in the form of biomass, hydro, solar, geothermal, and wind.

Oil and gas deposits are located in the Rio del Rey Basin, the Kribi region, the Benoue Basin of Logone Birni and in the Cameroon territory around Lake Chad. Cameroon has one oil refinery, in Limbe. The National Corporation of Petrol Stations (SDPC) manages national distribution of finished petroleum products. Substantial deposits of gas are located in Kribi and in the Garoua regions. Hydroelectricity is a principle source of energy in Cameroon; hydroelectric plants are located in Edea, Song Loulou and Lagdo.

Biomass is an extremely important energy source. The country possesses 20 million hectares of natural forest and 75 per cent of the country is classified as having a large potential for the production of wood and biomass. Some examples of how biomass is being used include: in the North Province, where cotton waste is used for energy production; in the Littoral Province, where palm kernel wastes are used for energy production; in the Centre and Littoral Provinces, where wastes from timber production are used to produce energy; in the Centre and Littoral Provinces, where sugarcane wastes are used for the production of energy.

The most favourable zone for solar energy is the Northern region of the country. Hundreds of small solar home systems are in place, and their numbers are growing. Local entrepreneurs are promoting these systems. Total installed solar power remains low, however.

Wind energy has great potential in Cameroon. In the Kaele region, for example, average wind speeds reach 4.5 metres per second for nine months in a year. Wind energy is, nevertheless, still only at the experimental stage, with pilot installations located in Pitoa, Maroua and the Moulvoudaye Centre.

Cameroon possesses great potential for the development of geothermal energy, and geothermal resources are distributed throughout the country. The Ngaoundere region, for example, has resources in Laoupouga, Katip and Foulbe, while the Manengouba region has resources in the region of Lake Monou. Nevertheless, Cameroon has not yet benefited from even one pilot or demonstration geothermal project.

Cameroon's national energy programme focuses mainly on developing regional integration of power from large hydroelectric plants, and on privatization. The programme pays no attention to the development of small-scale projects for rural areas, which are not connected to the electricity grid. The general economic crisis facing the country has led the government to propose measures to encourage energy conservation in state-owned companies and public services. These measures have been promising but have had little effect because of a general lack of interest in their implementation among administrative staff and a general lack of follow-up or monitoring.

The way in which privatization has taken place to date in Cameroon has hampered the development of small-scale hydroelectric power. This is because of a law, dating from 1983, that relates to government concessions for power production. Under present conditions, this law prevents private companies from taking over small-scale hydro schemes (of under 1,000 kW) that have been abandoned by the state electricity company. Meanwhile the state is still focusing mainly on large hydro development, as mentioned above.

In Cameroon there is no government regulation in place to define clearly the use of new and renewable energy, energy efficiency or integrated resource planning. Indeed, the rational use of energy is not covered by any of the country's laws. Meanwhile, high taxes and custom duties on solar photovoltaic products, for example, have the effect of discouraging their propagation and popularization. Furthermore, regulations governing access to information are problematic. Information concerning government energy studies, project designs, and implementation reports is released on a discretionary basis only, and without clear guidelines or criteria. This means that much information is simply not available to the public; it is therefore very difficult to carry out independent research into the country's energy sector.

Song Loulou Hydroelectric Plant

The Song Loulou hydroelectric power plant is the largest in Cameroon. The plant, on the Sanaga river in the Littoral province, is the site of a five million cubic metre storage reservoir and a power station housing six 48-megawatt generating sets, including one standby set, providing a peak capacity of 240 megawatts, and annual generation of 2,100 gigawatt hours. The estimated cost of the project is US$230 million. Work began in 1976 and ended in 1980. The plant was funded by a number of financial institutions: the European Investment Bank (US$15 million); Caisse Centrale de Cooperation Economique (US$43 million); Saudi Development Fund (US$30 million); Kuwaiti Development Fund (US$15 million); Qatar Government Fund (US$3 million); Islamic Development Bank (US$7.35 million); private banks supplied US$25 million, the National Electricity Corporation (SONEL) provided US$45 million, and other credits came to US$46.65 million. The project was controlled and supervised by the National Electricity Corporation.

The nearest town is Massock, with a population of about 12,000 inhabitants. The hydroelectric plant itself is situated in the heart of the equatorial forest. The area was uninhabited at the time that work on the project began. Local traders and construction employees started occupying the area after the commencement of construction works.

Expensive automatic equipment broke down soon after the plant became operational. Local engineers do not have the capacity to maintain and operate this equipment, so the plant is now being operated on a mechanical basis.

Of the five villages nearest the plant, only three are electrified. This is causing a steady exodus of younger villagers towards the towns. Also, tourism is restricted in the area; the National Electricity Corporation says that this is a security measure designed to safeguard the power plant.

The scheme has led to some deforestation, although this is not very significant given the vastness of the surrounding forest area. The project has, however, led to the presence of some harmful insects that cause various illnesses, such as ophthalmic diseases and malaria, because of areas of stagnant water that have been created near the dam. Initially, the National Electricity Corporation took regular steps to destroy these insects. In recent years, however, this vital programme has been suspended.

Today, the people in the surrounding villages are generally displeased with the plant. Roads are in disrepair and villagers are forbidden from fishing. Villagers further suffer from a general lack of basic facilities and services. There is no adequate health centre, for example, and supplies of clean drinking water are limited.

The Mape Dam

The town of Mape, with around 12,000 inhabitants, is situated on the River Mbam in Cameroon's West Province. The most important economic activity is agriculture. In addition, small-scale subsistence fishing and semi-industrial fishing are practised.

The Mape dam was constructed between 1984 and 1988 at a cost of US$67.05 million, funded mainly by the African Development Bank, and carried out by the Italian company Impreglio. The dam has the capacity to retain 3.3 billion cubic metres of water to reinforce the production of the Song Loulou dam during periods when water levels are low.

The dam project covers an area of 45 kilometres and has directly affected nine nearby villages. In addition, it has resulted in the disappearance of the natural vegetation in the flooded area, and has disrupted the Savannah ecosystem in the Highlands. The National Electricity Corporation promised compensation to local people affected by the project. Many local people questioned, however, have indicated that proper payment has not yet been made.

There has been an increase in ophthalmic diseases in the area. These are being transmitted by insect species that have benefited from areas of stagnant water created by the dam. Also, diseases such as bilharzia, malaria and scabies are rampant in the area. These health problems are being further multiplied by a serious lack of medical care, hospitals or dispensaries.

In addition there is a lack of educational services, clean drinking water and adequate road infrastructure. Moreover, electrification in the area is minimal and kerosene is most often used as an energy source for lighting. With access routes to the area being so difficult, however, supplies of kerosene are scarce and prices are consequently high.

No Effective Framework for Sustainable Energy

Cameroon's national energy policy is based on the principles of the preservation of energy independence and the development of external exchange; the promotion of access to energy at national level and competitive prices; the utilization of energy as an incentive to economic growth and employment; environmental preservation and the improved security of supply; and the improvement of legal, regulatory and institutional frameworks.

This national energy policy has been influenced by the energy policies of the World Bank, especially on the issue of competitive pricing. National energy policy differs from that of the Bank, however, on the issue of privatization. In fact, SONEL and the National Hydrocarbon Corporation (SNH) finance all their activities from the national budget. Although SONEL exercises a monopoly over hydroelectric energy, there has been partial privatization in the hydroelectric sub-sector, where projects have been contracted out to independent power producers. The oil and gas sectors, on the other hand, remain entirely state controlled and do not fulfil the conditions in the Bank's energy policies of transparency or privatization.

In relation to transparency, decisions on energy investments are not subjected to public scrutiny and do not involve local populations or even local government officials. Furthermore, although a law was passed in 1990 on liberty of association, which supported cooperation between government and NGOs, state bodies and parastatal enterprises are on the whole unwilling to release information to the public. Other regulations and laws relating to the release of information have, similarly, not been implemented. On the whole, there is no public involvement in energy planning in the country and there are no efforts by the World Bank or other MDBs to promote this.

Cameroon possesses a diverse climate, a rich geography and subterranean wealth, as well as a range of exploitable renewable and fossil energy sources. Nevertheless, the energy needs of the population are not being met. This lack of access to energy is most acute in the rural areas, where the vast majority of the population live. Rural areas have a largely unmet demand for energy. Wood is mostly used for cooking and space heating, while kerosene is mostly used for lighting. Rural energy development is a low priority for energy investments, however. Rural electrification schemes have been carried out for political reasons with very little accountability. In general, there is not enough finance made available for investments in rural energy. There is no rural energy scheme for the country and fewer than 5 per cent of people in rural areas have access to electricity.

Solar home systems have gone some way to meeting electrification needs in rural areas. Nevertheless, the market penetration of such systems remains low. This is due partly to lack of promotion within Cameroon itself, and partly to national fiscal and customs policies that constrain the import of the mostly foreign equipment.

In fact, there are no examples at all within Cameroon of an investment in a major renewable energy project. This is the case not only for solar home systems, but also for wind, biomass and biogas energy. The activities of the World Bank in Cameroon have focused on trying to promote macro-economic reform in the direction of commercialization and privatization and on competitive energy pricing. The Bank has paid lip-service to renewable energy and rural energy in Cameroon, but has not backed up this rhetoric with investments. Meanwhile there has been little or no attention paid to the creation of an effective legal or institutional framework within the country to promote solar, wind, geothermal or small-scale hydropower, or energy efficiency.

Climate change is not an issue in government energy planning or part of any direct MDB activities in Cameroon. The World Bank's investments in the country, on the contrary, are focusing on the promotion of fossil fuels. The Bank is investing, for example, in the Chad/Cameroon Oil and Pipeline project, which will develop the Doba oil fields in neighbouring southern Chad and build a 600-mile pipeline through Cameroon to an Atlantic port for export. The predicted negative social and environmental impacts of the project have already become the focus of an international NGO campaign involving independent human rights and environmental organizations throughout the world.

Cameroon is in fact caught between, on the one hand, the inability of the state government to promote environmental protection and sustainable social development, and, on the other hand, a lack of commitment from the World Bank and other MDBs to policies on rural energy and the promotion of renewables and energy efficiency.

Note

1. Analysis of Cameroon prepared by Jean Koueda Koung, Global Village Cameroon, B.P. Yaounde, Cameroon.

4

Kenya[1]

A Deteriorating Situation

In 1991 the World Bank imposed a freeze on international development aid to Kenya's energy sector. This was partly in response to the Kenyan government's refusal to implement energy sector reforms – including those related to private sector involvement. As a result, investments in generating capacity were halted, causing power outages, load-shedding, rationing and blackouts. The Kenyan government has estimated that there is a need for investment requirements of about US$1 billion over the next five years to meet the growing demand for electricity.

Against the background of a deteriorating situation in the energy sector, the government of Kenya agreed to undertake the energy sector reforms that have been laid down by the World Bank. The aid embargo on the energy sector has been lifted, and in November 1996 the World Bank agreed to lend Kenya US$125 million as part of a US$700 million package of investments in the country's electric power sector.

To date, the government has invited bids for independent power producers to produce power in Kenya. It has also recently restructured the power sector to create separate generation, transmission and distribution companies, operating independently of each other. It has not yet taken the necessary steps, however, to create appropriate legislation on the regulation of the sector.

Kenya's economy depends on three major sources of energy: fuelwood, petroleum fuels and electricity. Minor energy sources in the country include solar, wind, ethanol, coal and biogas.

Fuelwood meets approximately 70–80 per cent of the country's domestic energy requirement and is largely consumed in the rural areas and as charcoal in urban areas. Industry takes as much as 26 per cent of fuelwood and 12 per cent of charcoal consumption. Kenya is thus heavily dependent on a declining resource (fuelwood) for household energy needs and on imported oil for other energy requirements.

Imported crude and petroleum products are most crucial for the industrial, transport and agricultural sectors but have also penetrated all major socio-economic sectors, including rural households. Petroleum fuels (and electricity) are currently the major sources of commercial energy and their adequate supply is crucial to the country's current development strategy. Imported petroleum fuels currently account for 67 per cent of the country's total consumption of industrial and commercial energy.

Kenya's electric power is generated from hydro, thermal and geothermal plants. Electricity plays a crucial role in the country's development process, being a key input in various industrial processes, although only 14 per cent of Kenyan households are consumers of this form of energy. The urban bias of electricity use is so strong that rural households have only a 1 per cent share of total electricity consumption. Kenya's grid thus serves industry, residents and commercial customers in urban areas almost exclusively (Hankins 1993).

Thermal- and geothermal-based power generation is, however, not well developed, although Kenya is regarded as one of the few developing countries that have undertaken a significant development of geothermal energy resources. Hydropower is by far the most important source of commercial primary energy currently produced in Kenya. The combined total installed electricity capacity inclusive of thermal and geothermal is about 842 MW.

Over the last ten years, the installed electric capacity has grown rapidly and current official estimates indicate that demand is projected to grow at an average of almost 5 per cent per year. Even though the capacity of the sector has grown, however, the proportion of the population served has increased only modestly and is currently estimated to be less than 8 per cent of the population.

Kenya's major energy consumers fall into five broad sectors: household, commercial, manufacturing, agricultural, and transport. Of these, the household sector accounts for about 59 per cent of national energy use, followed by transport (13 per cent), manufacturing (15 per cent), agriculture (the backbone of Kenyan economy) (9 per cent), and commercial (5 per cent). Envisaged economic and population growth combine to create a virtually limitless demand for power services.

Energy conservation in the country has not been promoted to any great extent. While the need for conservation of energy became apparent from the time of the 1973/74 oil crisis, effective conservation has been hampered by a lack of general conservation awareness by the population; old, insufficient machinery and equipment; lack of information on energy efficient practices, equipment and appliances; lack of adequate standards and regulations that take into account energy conservation codes and practices; and the high cost involved in replacing machinery. Additionally, energy conservation is constrained by regulatory measures such as energy pricing and fiscal policy that tend to constrain its development.

The Ministry of Energy operates two major energy efficiency programmes – the Kenya Energy Management Programme (KEMP) and the Kenya Energy Auditing Programme (KEAP). The former, which is administratively under the Kenya Association of Manufacturers (KAM), deals basically with the dissemination of information through seminars, publications and visits on benefits of conservation measures while the latter, which is the first initiative to institutionalize national energy auditing efforts, analyses patterns of energy use and distribution and identifies opportunities for energy saving. Both the World Bank and the United Nations Development Programme also support energy conservation, particularly in the industrial sector through the Energy Sector Assistance Programme (ESMAP).

About 70-75 per cent of Kenya's population is rural based. Like most of Africa, Kenya's energy sector can be divided into a modern commercial sector and a subsistence rural sector. The rural subsistence sector demands the highest proportion of the country's energy supply in the form of wood and charcoal for cooking and space heating as well as kerosene (known in Kenya as paraffin) for lighting.

Despite rhetoric in policy pronouncements, the development of rural energy in Kenya is not a priority. The Rural Electrification Programme, for example, which began in 1973, operated several schemes successfully until around 1988. During this period power-lines were extended to all district headquarters throughout the country. Since 1988 there have been claims that rural electrification schemes have been undertaken largely on a political basis, however, with little accountability. Twenty million Kenyans, out of a total population of 26 million, live in rural areas. Less than 1 per cent of this rural population has direct access to electricity, however. The main reason for lack of progress in rural electrification has been lack of investments needed to meet the high capital costs for the extension of the power grid.

The government bureacracy in Kenya has not engaged to any great extent in the development and management of biomass (wood, charcoal, agricultural wastes, biogas), solar, wind, power alcohol or micro-hydro. This is partly because these energy sources do not provide much in terms of immediate political or economic returns. There has been little meaningful follow-up on successful pilot schemes, such as the ceramic 'jiko' improved woodstove. In addition, solar equipment continues to be highly taxed, which hinders its promotion, maintaining it as a preserve of a select privileged few.

Available data on the wind regime in the country still do not provide a sufficiently sound basis for large-scale investment decisions, while solar heating at today's electricity tariffs and fuel oil prices is not financially attractive. Nevertheless, these sources of energy have the potential to supply large amounts of energy and can be used over much of the country to supplement conventional sources.

Solar energy is today the most rapidly growing non-conventional source of energy in Kenya and is most commonly used in water heating in homes, small hotels and institutions such as especially schools and colleges. Photo-voltaics (PV) in Kenya has now been developed by a number of individuals on a commercial basis. So far over 20,000 systems have been installed in scattered places throughout the country and there is potential for more of the same. It is estimated that today more rural households have solar electricity than grid electricity, despite the fact that grid electricity is subsidized while solar equipment is highly taxed.

Wind has been known and exploited as a primary source of energy in Kenya for well over a century, and has mainly been used for sailing, milling and water-pumping. Although the tropical winds, particularly those close to the Equator, are generally light, there are many potential sites within mountainous areas, coastal strips, lake-shores and other well-exposed areas where wind energy can be harnessed. Nevertheless, there has been little effort by policy-makers or implementing agencies to popularize wind-powered technologies in the country. This is despite the fact that the potential for wind energy to contribute to the national energy supply is likely to be significant.

The development of power alcohol in Kenya started in the mid-1970s as part of the government strategy to overcome dependence on imported petroleum products by replacing imported fuels with domestically pro-duced fuels. This strategy also included the production of seed oils from energy crops and fast-growing trees. Little attention has been given to this source of energy, however, and the blending of alcohol with gasoline currently accounts for only 10 per cent of total engine fuel, with most of this being consumed in Nairobi and the surrounding towns. With proper planning and adequate investment, power alcohol has great potential as a source of energy in Kenya, especially given the vast sugar-cane plantations in the Western Kenya region.

Small-scale hydropower (micro-hydro) is yet another source of energy the potential of which has not been harnessed. There are several scattered sites, mainly in the western part of Kenya on the Yala, Nzoia, Arror, Kuja, Sondu and Miriu rivers. The priority of the government and development agencies is nevertheless still focused on large-scale hydropower projects and the micro-hydro sector remains undeveloped.

The Energy Sector Reform and Power Development Project

The Energy Sector Reform and Power Development Project is the first major World Bank funded project in Kenya's energy sector since the adoption of the Bank's new energy policies in 1992. The total cost of the project is relatively large at US$699.9 million. The World Bank's share of the investment is just under 18 per cent, at US$125 million. The Bank is providing this loan under its soft-loan window, the International Development Agency (IDA). Other financiers are the Overseas Economic Cooperation Fund of Japan (OECF) (about US$82.8 million); the European Investment Bank (EIB) (about US$48.7 million); Kreditanstalt für Wiederaufbau (KfW) (about US$20.8 million) and private sector investors (about US$262.5 million). Kenya Power and Light Company (KPLC) and Kenya Power Company (KPC) will finance the balance of US$160.1 million and the charge of US$99 million for interest during construction. The project involves five new power plants with a combined capacity of 338 MW during the period 1997–2001. These are Kipevu I Diesel Plant (75 MW) Ol Karia II Geothermal Plant (64 MW) Sondu-Miriu Hydro Plant (60 MW) Kipevu II Diesel Plant (75 MW) and Ol Karia III Geothermal Plant (64 MW). The project has six main components:

1. Sector restructuring and reform: estimated to cost US$24.6 million, the component would include: (a) support for establishment of legal and regulatory framework necessary to improve sector efficiency; (b) reform of the organization, management and financial structure of the power sub-sector companies, and separation of generation from transmission and distribution functions; and (c) promotion of private sector participation in the provision and management of operations.

2. Institutional support: this will cost US$24.6 million and comprises studies, advisory services and logistical support to project implementing entities.

3. Efficiency improvements: estimated to cost US$11.8 million and has two components. First, demand-side improvements that include developing capacity in KPLC and in the local private sector to design, implement and evaluate efficiency and electricity demand management projects. Second, line loss reduction that will finance major distribution rehabilitation and loss reduction programmes in the Nairobi and coastal areas.

4. Power system expansion and rehabilitation: this will cost US$609.1 million and has two components: (a) to increase generation capacity and to install associated transmission facilities; and (b) a programme for the reinforcement of the primary distribution systems in Nairobi and in the coastal areas.

5. Geothermal resource development: this is estimated to cost US$49.3

million and would assist the government in developing geothermal energy resources for future private sector participation.

6. Future project preparation: estimated to cost US$1.5 million, the component would support preparatory activities for the follow-on projects.

From a general perspective, the Energy Reform and Power Project can be said to be only partially in compliance with the Bank's 1992 energy policies. This is because although the project is based on integrated energy strategies, these are only partially supported. The project documents mention efficiency improvements on the supply side, such as line loss reduction. Further, sector restructuring and reform – meaning commercialization and competitive pricing – are included as set measures to increase general efficiency. Crucial aspects of integrated energy strategies, however, such as energy conservation, are not emphasized. Further, despite the importance of women in the household use of biomass in Kenya, the project makes no mention at all of gender issues. In general, it puts most of its emphasis on privatization, supply-side management, the construction of new power plants, the rehabilitation of existing equipment, transmission and distribution system expansion, and the institutional development of utilities along commercial lines.

Since the project has not been fully implemented at the time of writing it is still too early to assess the degree of compliance or non-compliance with environmental assessment requirements. One major omission that can already be seen at this stage, however, is that environmental and social externalities and the associated mitigation costs of the thermal and hydropower components are not specified in the project design.

The project does include one demand-side management component (see Box 2.3), which involves developing the capacity of the Kenya Power and Light Company and the local private sector to design, implement and evaluate efficiency and electricity demand management. This DSM component makes up only 1.6 per cent of the total investment, however, while more than 87 per cent of the loan is intended for power system expansion and rehabilitation, comprising mainly new diesel, hydroelectric and geothermal power plants. The environmental assessments carried out on the project to date have indicated the following major impacts:

- The site of a planned 32-MW geothermal plant at Olkaria is rich in wildlife and serves as a grazing area and migration route for Maasai pastoralists. The project has potentially serious pollution problems arising from the discharge of steam and hot water with high mineral content. The project will mainly affect wildlife and livestock, and lead to soil erosion and loss of habitat for large herbivores. In addition, the project will entail limited resettlement of local people.

- A 75-MW diesel plant at Kipevu will lead to hazardous waste dumping, oil, silt and heavy metal run-off, and downhill siltation.
- A 60-MW hydroelectric power project at Sondu Miriu will divert water from the Sondu river, leading to disruption of the fragile river eco-system, loss of agricultural and residential land, water and soil pollution during construction, and disturbance to the Koguta forest.

Environmental mitigation plans have been drawn up to deal with each of the above impacts. Kenya Power Company's Environment Unit will monitor mitigation. But neither Kenya Power Company nor Kenya Power and Lighting Company have the capacity to carry out such work, so staff training in environmental analysis and impact monitoring has been planned.

The World Bank's energy policies mention the importance of NGO and community participation in project design and implementation, although there is little evidence of a participative approach in this project. Local communities were involved in the project only while initial studies were being carried out. 'Participation' at that time involved no more than providing data to the researchers working on the feasibility studies.

Some components of the project are clear violations of existing international agreements. The geothermal project at Olkaria, for example, will disturb wildlife habitats and vegetation in violation of the UN Convention on Biodiversity. There is no evidence, furthermore, that the project has taken on board the people-centred development approach that was emphasized during the UN Social Summit on Development. And, importantly, two diesel-powered generation plants being developed in Mombasa and Nairobi South will violate the spirit of the UN Framework Convention on Climate Change and the Kyoto Protocol (see Box 2.4).

Gaps Between Policy and Practice

Since the 1970s, the World Bank has been the major MDB and lead financier in Kenya's energy sector. Other MDBs, such as the African Development Bank and the European Investment Bank, have not yet become significantly involved. Until the 1990 aid embargo on Kenya the Bank had invested large amounts for the development of mainly hydro-power plants and geothermal energy. These investments were all made before the adoption of the Bank's 1992 energy policies, however.

The Bank's 1992 policies put forward principles relating to energy efficiency and on 'meaningful progress towards transparent regulatory processes'. It is clear that in Kenya there are gaps between these policy statements and actual practice. The above examination of the Energy Sector Reform Project, for example, shows that the Bank's emphasis is still on financing large scale hydropower, geothermal projects and fossil fuel projects, despite the well-documented negative environmental and economic impacts of these. The Bank's commitment to the development of non-conventional sources of energy in Kenya is weak, and it is giving much more attention to supply-side efficiency measures (such as reducing power lost from inefficient power-lines) than to demand-side management (see Box 2.3).

The Energy Sector Reform Project is, further, a clear example of the World Bank applying pressure on a low-income country to commercialize and privatize its energy sector. It can be argued that the privatization of the energy sector will provide incentives for increased energy efficiency and improve on public sector mismanagement by encouraging competition and removing the monopoly enjoyed by major sectoral players. In Kenya, however, independent power producers (IPPs), looking for profit maximization, are turning to conventional power projects before investing in new and renewable energy technologies or in demand-side management programmes. It is further likely that, given their profit motivation, the policies of private entrepreneurs will be at variance with government public sector lending standards in the areas of environmental assessment, participation, and information access and disclosure. Rural areas are also likely to lose out most under a regime of solely private power production, as private companies will be reluctant to enter into less attractive investments in remote rural areas of Kenya. The government's already weak ability to carry out environmental management is likely to suffer under privatization of the Kenyan power sector. Privatization is an important measure with a substantial potential, but it is being implemented in Kenya as a panacea, and this is not producing the intended results.

The Energy Sector Reform Project also demonstrates that despite the overwhelming economic and environmental advantages of end-use efficiency and conservation investments, the Bank is concentrating on large-scale, capital-intensive energy expansion projects.

In addition, the Bank does not seem to have addressed the issue of renewable energy. Kenya faces the linked problems of deforestation and scarcity of fuelwood – the main energy source for the country's households. If present trends (including 3.4 per cent annual population growth) continue, annual yields can meet only 11 per cent of fuelwood demand in the year 2000; stock would be able to supply only a further 24 per cent, indicating an actual shortfall of 65 per cent of national fuelwood demand. Indeed, the high percentage of people dependent on fuelwood for cooking is environmentally and economically unsustainable. This high dependence is damaging to soils, forests, agriculture and people's health. Fuelwood is also inefficient compared to electricity or liquid petroleum gas.

Also, the issue of greenhouse gas emissions is of global concern. Energy is the sector that most profoundly contributes to these emissions. The Bank's main investment in Kenya since its 1992 energy policies is biased towards oil-fired power stations and does not help in the global fight against climate change. On the contrary, it is evident that the Bank's lending priorities for energy development and infrastructure in Kenya remain skewed in favour of fossil fuels.

It is now widely accepted that social factors and local participation play a significant role, particularly in rural energy development. The success of renewable energy projects depends principally on the effectiveness of local community mobilization, participation and organizational skills. While the Bank's policies make commitments to participatory strategies within its investment projects, there is little, if any, evidence of community or NGO participation in its energy investments to date in Kenya.

Note

1. Analysis of Kenya provided by Hudson Isagi, Resource Projects Kenya (RPK), Nairobi, Kenya, P.O. Box 76406, Nairobi, Kenya.

5

Zimbabwe[1]

Energy and Development in Zimbabwe

Zimbabwe does not have a clearly articulated energy policy. Pronouncements have been made about the need to provide adequate energy supplies to all sectors of the economy, but the only energy sub-sector apparently guided by a government instrument is the electricity sector. The interlinkage of the various sub-sectors in energy and their final impact on the economy require that a comprehensive energy policy be in place. This is especially imperative for the rural areas, where years of neglect have left a virtual energy crisis time-bomb. Government evaluations have looked at the major commercial forms of energy – electricity, coal and liquid fuels – almost to the total exclusion of all other forms of energy, including biomass. The focus on commercial fuels serves to highlight the utility of these forms as opposed to the traditional forms of energy. Fuelwood and other renewables have received very little attention from a policy perspective in Zimbabwe, mainly on account of a failure to attach an official market price to the resource. Such a position is misleading considering the environmental costs that result from deforestation and soil erosion, which ultimately end up being paid for by the economy as a whole.

The Ministry of Transport and Energy (MTE) has overall responsibility for the energy sector in Zimbabwe. The Ministry's Department of Energy is responsible for the planning and coordination of activities in the commercial fuels sub-sector, but also has an interest in the domestic fuels. Three major parastatals fall under this ministry: the Zimbabwe Electricity Supply Authority for the provision of electricity, the Zambezi River Authority, which is responsible for the development of hydropower in a section of the Zambezi river owned by Zambia and Zimbabwe, and the National Oil Company of Zimbabwe, which is responsible for all bulk liquid fuels procurement. Besides the MTE, the Ministry of Mines is also involved in the energy sector through its control over mining and exploration for hydrocarbons in the Zambezi valley. Coal-mining in Hwange is through the

Wankie Colliery Company, in which the government owns shares. The Forestry Commission, a parastatal in the Ministry of Environment and Tourism, oversees the supply and utilization of the country's woodland resources. The private sector is also involved in the energy sector, especially in the distribution and marketing of liquid fuels. Efforts are under way to develop the extensive coal reserves in the Sengwa area and to exploit natural gas in Matabeleland North province. With the launching of the GEF–PV project, which seeks to supply photovoltaic systems to rural households, a number of firms in the private sector have begun to electrify rural areas and some firms have started manufacturing system components. The manufacturing capacity of local firms is likely to grow with the increase in PV installations. Decisions on the pricing of coal, electricity and petroleum products in Zimbabwe are the responsibility of government through the Cabinet Committee on Development. The Ministry of Finance is responsible for securing external finances for the funding of investment projects in the energy sector and the economy at large. The government therefore largely controls the energy sector through the influence its institutions have on energy planning, pricing and investments.

A dimension of the energy sector that is evolving rapidly is the transboundary trade in electricity with the various utilities in the region forming the Southern Africa Power Pool. This brings the Southern African Development Community (SADC) Energy Sector Technical Advisory Unit in Angola more to the forefront as a regional energy body. There are more than eighteen NGOs involved in the energy sector in Zimbabwe, conducting research, disseminating information, providing training courses and implementing energy projects in rural areas.

A Considerable Energy Endowment

Zimbabwe has a considerable energy endowment consisting primarily of biomass, coal, hydropower and solar energy. Liquid fuels are imported and coal resources are found in 21 fields spread throughout the country, with an estimated potential reserve of up to 10.6 billion metric tonnes. Almost all the country's hydropower potential is in the Zambezi Basin, which Zimbabwe shares with Zambia. The total hydroelectric potential in the Zambezi region is estimated at 37 TWh per annum, of which roughly one-third has been developed at Kariba and Victoria Falls. A total capacity of 666 MW has been installed at Kariba South, while 600 MW has been installed at Kariba North. Upgrading of Kariba South will raise the capacity to 750 MW. In addition, new projects are being investigated on the Zambezi and these include the Batoka Gorge hydroelectric scheme (1,600 MW) and the Kariba South extension. Very little other hydro potential exists in the country except for a few mini-hydro sites in the Eastern Highlands. This substantial infrastructure serves only 20 per cent of the population, that is, the urban areas and commercial farms. Biomass in its various forms has played an important role in meeting energy needs and is available from 7.5 million hectares of accessible forest land across 20 per cent of the country. In most of the country fuelwood is considered a free resource. The burden is especially severe for women, whose task it is to collect fuelwood for the household. Fuelwood supplies about 26 per cent of Zimbabwe's total energy demand. There are roughly three categories of users and consumption is distributed between them as follows:

- communal areas 74 per cent (or 4.7 million m³);
- urban households 16 per cent (or 1 million m³); and
- commercial agriculture 11 per cent (or 0.7 million m³).

Alternatives to Environmental Damage

Environmental degradation is evident in many of the communal areas, with rural industries contributing significant impacts. The greatest proportion of the damage stems from the destruction of natural forest without a corresponding replacement of these lost trees. With the removal of forest cover sheet erosion results from the action of the wind and rainfall after a loss of vegetation cover. The haulage of logs along the ground or on sleighs contributes to the formation of gullies along the tracks, which then provide a channel for erosive rainwater. If the land is left in this state year after year much topsoil is lost. Tree planting of either indigenous or exotic species has not kept pace with the rate of tree loss. Efforts continue by the Forestry Commission, the extension agencies and NGOs to promote tree planting, but the pace is slow.

Concerted efforts have been made by the Department of Energy to bring renewable sources of energy into the spotlight, with various initiatives on biogas and solar being proposed. The funding of the US$7.5 million 'Photovoltaics for Household and Community Use Project' by the United Nations Development Programme (UNDP) added impetus to the thrust towards solar as an energy source worth investing in. Enthusiasm for solar can be understood since there is relatively high solar energy potential in the country, with daily radiation estimated at 2 kilojoules per square centimetre and most parts of the country receiving approximately 3,000 hours of sunshine per year. Very little work has been done by researchers on the issue of technology in the use of fuelwood, but it has emerged that the technologies to utilize the fuelwood resource efficiently are subject to the same barriers as other improved technologies that could improve economic prospects. The first barrier to technological improvement in this sector is that fuelwood is still regarded as free, a common access resource without the need for efficiency measures. Attaching a monetary value to it will prompt action. The fuelwood sub-sector suffers from a lack of policy attention and Zimbabwe's own energy planning process does not take into consideration the fact that 80 per cent of people rely to varying degrees on the use of fuelwood for energy. Biogas can be used for lighting and electricity generation for both domestic and commercial purposes. Residue from the process can be used in agriculture to enrich the soil and to support such activities as aquaculture, mushroom production and stock feed. When linked to sanitation devices it destroys bacterial pathogens and enteric viruses and makes for better health in the community. An added advantage of the technology is that it provides opportunities for diversification of economic activity in the areas. Despite these advantages, the technology has not been fully explored in Zimbabwe. To enhance the dissemination of biogas technologies it is also necessary to develop the manufacturing component so that many more suppliers of gas equipment are available. There is much debate as to whether the structure for biogas dissemination should be centralized or decentralized. Further, a significant volume of biomass waste is generated in the country, especially in the sawmills in the eastern highlands and in Matabeleland North. Sawmill waste and coffee husks make for some of the best inputs in the gasification process. Agricultural waste such as cobs from large estates can also be useful. The process, however, is technical and needs a very accurate synchronization of various components in order to function properly. The high cost of the technology is another factor hindering the adoption of gasification plants. Several companies in Harare operate coke gasifiers, which are more efficient and less problematic than the biomass variety. These units produce up to 20 GJ of gas per hour. None of the companies has, however, measured the energy savings derived from the gas, but the installed units are very expensive. Ethanol is produced in the country's 'lowveld' from sugarcane and the maximum capacity of the ethanol-producing plant is 40 million litres per annum. Ethanol and fuel mixing is ideal for the running of stationery engine generators – a viable option for electrification in remote areas. The electricity generated is used both onsite and in the surrounding community. In fact, during the sugar season (April–November) the estate is virtually self-sufficient, with surpluses being sold to Zimbabwe Electricity Supply Authority (ZESA). Outside the season ZESA provides some 6 MW of power to the estate. The production of ethanol for blending with petrol has saved the country several millions in foreign exchange, and has enabled an entire community to be largely self-sufficient in its power requirements. Solar thermal and photovoltaic systems are promising in Zimbabwe, with several projects in place, some of which have been taken up by the commercial sector for widespread dissemination. These technologies, especially the former, have a great potential for rural areas, but their cost still makes them inaccessible to the majority of people in rural areas. Solar photovoltaics are suitable for such low-power applications as lighting, battery-charging, telecommunications, remote controls and refrigeration. The cost is still prohibitive for a majority of people, but it is hoped that with growing use of the technology and the utilization of economies of scale, costs can be brought down to manageable levels. Already there are more than three thousand installations throughout the country. Solar PV energy has encountered a number of problems such as affordability, so that it is mostly those on higher incomes, who in most cases can afford grid electricity, who have purchased it. Since the majority of households in the rural areas do not have access to grid electricity, however, PV has become the next best option. There is also the problem of standardization of systems, and lack of training/education has also been a significant factor in PV technology. This applies particularly to information relating to the repair and maintenance of PV systems and is important, especially since very few PV installers actually offer after sales back-up and some systems are non-functioning because of a relatively small technical fault, which could be easily repaired.

Mini-hydro is a very site-specific technology, albeit a very versatile one, but it has tended to be limited to the Eastern Highlands of the country. So far the Intermediate Technology Development Group (ITDG), an NGO, has done the most work on mini-hydro. It has installed systems in Nyafaru and Cashel, and has been involved in the Rusitu Power Corporation's 700-kW scheme, and the proposed 400-kW Manyuchi Dam scheme. Wind energy has limited applications in Zimbabwe due to the relatively low wind speeds with an average of 3 metres per second. Despite this limitation a few wind systems have been installed at several locations in the Eastern Highlands. Wind speed data that have been collected in several areas indicate that there is potential to install systems of between 250 and 750 MW.

The opening up of the economy and the emphasis by government on economic efficiency in Zimbabwe's parastatals has sent clear signals to the energy sector to consider improved efficiency in its operations. The dominant objective of economic efficiency and growth implies that sufficient energy at least cost has to be supplied to accommodate economic growth. Further, energy supply and demand management must be achieved in an environmentally sustainable manner. As a general principle the government of Zimbabwe is committed to creating an enabling environment for the attainment of economic efficiency in parastatals, and the economy as a whole is playing a facilitator's role.

The Department of Energy's mission statement is: 'To regulate the energy industry and facilitate and mobilize private and public sector participation in order to ensure adequate and affordable energy on a sustainable basis so as to promote economic growth.' The goals of this mission statement are realized through activities carried out in the areas of renewable sources of energy, power/electricity, petroleum fuels and energy conservation.

The Research and Development (R&D) section of the Department of Energy has the mandate to supply alternative sources of energy for rural development and it is committed to the sustainable supply of energy for rural development and household consumption through the promotion of alternative energy technologies such as biogas for lighting, cooking and refrigeration; coal for domestic and institutional use; solar PV for lighting, water-pumping and powering of small domestic appliances; solar thermal technology for outdoor cooking and water-heating; and wood-stoves, to promote fuel efficiency and improvement of the kitchen environment. As a way to fulfil this mandate the R&D section does a lot of work in the above specific projects.

The Bank's Weak Link to Environmental Issues

The World Bank's energy sector investment policies have been based on creating conditions that ensure that the client countries are able to pay back their loans. Therefore the emphasis of these policies is on energy efficiency, conservation and creation of an enabling environment and infra-structure to ensure that the project provides a good return. If anything the link to environmental issues has been rather weak. The following general principles guide the Bank's energy sector investments in Kenya: i) countries should have an integrated energy strategy in which energy efficiency and conservation are an important component; ii) developing country govern-ments must set policies and formulate strategies that point the energy sector towards the most efficient, equitable and environmentally compatible resource use that is feasible. It must be noted, however, that the emphasis on environmental aspects is a recent development; early energy sector investments did not stress environmental considerations.

The highest priority for improving the efficiency of energy supply and end-use in many developing countries must be to improve the basic institu-tional and efficiency-incentive structure relating to energy. On the supply side, the highest priority would be to make energy supply enterprises responsive through institutional and regulatory reform and increased private sector participation as well as to target plant rehabilitation, reduced transmission and distribution losses. Other priorities, such as facilitating the ease of technology transfer and increasing the focus on energy efficiency in transport, would involve initiatives on both the supply and demand sides. In addressing these priorities, the government must think long term and maintain consistent policies so that its strategic sector objectives will ultimately be met.

The creation of an institutional framework to encourage efficient energy production and distribution must include initiatives on at least two levels:

- restructuring energy supply enterprises; and
- facilitating a transparent regulatory mechanism between government and the energy supply enterprises.

The Bank has pledged its commitment towards increased lending for components designed to improve energy efficiency and to promote economically justified fuel switching. Since independence, the Bank has funded some projects in Zimbabwe that are entirely devoted to energy efficiency. Most of these projects have directly instituted technical changes and improvements, promoted energy conservation in all sectors of the country's economy, rehabilitated power and industrial facilities, and promoted demand and load management in the power sector. The Bank has also shifted its emphasis from investing in large hydro schemes and power stations to smaller plants such as mini-hydro and renewables. This is intended to improve the efficiency of the plants and address the socio-economic and environmental concerns of the recipient countries. From discussions held with Department of Energy senior officials and ZESA, it is apparent that the other MDBs that have invested in the energy sector generally follow the World Bank policy on energy sector investments. Although they may not be as strict as the World Bank in ensuring that their policies and conditions are adhered to, the general trend in their policy guidelines are based on the World Bank policy.

The biomass project As an alternative to fuelwood and paraffin for household use, biomass use can help protect existing wood stocks. The aims have also been expanded to the stabilization of sewage waste for institutions. Demonstration digesters at institutions have been built, as well as domestic digesters for individual households in Zimbabwe's rural areas. Biogas has also been used as a high-quality organic fertilizer for home gardens.

The coal stove project Many rural institutions, such as secondary schools, hospitals and other government institutions, use a lot of wood for cooking and water heating. It is common knowledge that close to 95 per cent of households use fuelwood for cooking and water heating. The

Department of Energy has therefore embarked on a programme to sub-stitute fuelwood with coal in these institutions as well as in households where coal could be made available at competitive prices, particularly in areas where there are wood deficits.

Solar PV In a bid to raise the standard of living of the rural people and to provide some basic facilities such as light to enable reading at night in schools and households, the improvement of health facilities and the supply of clean drinking water, the Department of Energy has promoted the use of solar energy for lighting, water-pumping and, in limited cases, refrigeration. As a result of this initiative a number of rural schools and clinics now have clean reliable supplies of drinking water and lighting and refrigeration facilities.

The petroleum sector In this sector the guiding principles aim at ensuring efficient and adequate supplies of fuel at affordable prices. This has been achieved through the formation of a single national buyer, the National Oil Company of Zimbabwe (NOCZIM), to take advantage of economies of scale and best market prices. Construction of strategic storage facilities ensures continuous availability of strategic stocks and putting into place a pricing formula that ensures viability of the sector. Increased availability in rural areas is achieved through promotion of the establishment of service stations at viable points. Adequate supplies are ensured by the fact that policy guidelines specify that stock levels are maintained at 60 days. The construction of the pumping-line from Ferruka to Harare and the strategic storage facilities enhance the achievement of this objective as well.

NOCZIM, which is the national buyer for petroleum fuels, awards tenders to the most competitive bidders for various products and then sells to the wholesalers, which are the private companies involved in the oil industries such as British Petroleum (BP), Shell, Mobil and Caltex.

Pricing is based on the Harare landed price, which constitutes import and transportation costs. On this landed price, a mark-up – which is calculated according to the return based on pre-agreed assets – is added for both NOCZIM and the wholesalers; this ensures viability of the in-dustry. Government duties and levies are also added. Efficiency in the industry, which is assured of returns by virtue of the pricing formula, is encouraged by use of an agreed efficiency factor that is built into the inflation adjustment figure. The more efficient the wholesaler, the more able it is to absorb the inflation effects. There is product cross-subsidization in the petroleum sector. The objective is to ensure that products that benefit agriculture, such as diesel and paraffin, which are used by people on low incomes, are kept at affordable prices (which are lower than their actual costs).

There has been a discernible increase in environmental awareness and the inclusion of environmental considerations in energy investment projects that have been approved and funded by the Bank. For instance, in contrast to the past, the Bank is now seen to be stricter on environmental impact assessment studies before it approves funding. In analysing the investments that have been funded by the World Bank in Zimbabwe, however, it can be seen that the Bank has tended to concentrate on funding conventional energy systems rather than renewable energy technologies. Emphasis on renewable energy has been fragmented, with a deliberate bias towards renewable energy technologies that promote the commercial interests of developed countries rather than recipient countries. An example is how the Bank was more interested in promoting solar energy technologies produced in industrialized countries than biogas technologies that suit developing countries, from both a financial and ecological point of view.

As already mentioned, however, it should be noted that the environmental protection guidelines are based on concern for the local environment as opposed to an all-out effort to embrace global climate change issues. If issues of climate change have been addressed in these investment projects then this has been as a spin-off from local concerns. The basic local concerns are the preservation, restoration and enhancement of the quality of ambient air and the environment as a whole for optimum social and economic use. Implementation of these controls has resulted in spin-offs such as reduced emissions of nitrous oxides (NOX) and sulphur oxides (SOX). Although the levels may be small by virtue of the scale of production they nevertheless do make a contribution towards the reduction of greenhouse gases (GHG). It is interesting to note that a monitoring station has been set up at the Hwange power station to collect data on wind speed and direction, ambient air temperature and rainfall, while there are no measures being taken to measure GHG emissions. This is an indication that concerns are focused on the local level rather than on climate change issues.

MDB energy sector investment policies are fairly consistent with national energy sector investment policy as far as environmental issues are concerned. It is, however, sad to note that there are some inconsistencies in policy and thinking when it comes to the type of investment to be made. Zimbabwe would like to build capacity in its local industries, and initial affordable finance is needed to build capacity. The Zimbabwe and World Bank policies differ here, however. Although the Bank's policy is shifting from funding large-scale commercial power plants to funding renewables, the objectives being to address environmental sustainability in supply and use of energy and socio-economic issues, it will still not fund at concessionary rates. The Bank has agreed to assist the Solar Energy Industries Association of Zimbabwe but has also insisted that the money cannot be lent at concessionary rates. There is an inconsistency here since loans for educational and health development can be obtained at concessionary rates, while those for capacity building within the energy sector cannot. While it is understood that the Bank should not give grants for commercial activities, it is difficult to separate commercial and social interests in the case of small-scale energy projects such as the promotion of renewables, and creating the necessary capacity in this industry.

An analysis of MDB investments, particularly those of the World Bank, clearly shows a bias towards efficiency improvement in planning, through the creation of ZESA; the introduction of Management Information Systems and Long Run Marginal Costing Studies; power generation efficiency improvement (thus rehabilitating old power stations); upgrading Kariba South; structural reform through the introduction of performance related rewards; and the pending privatization of ZESA. Zimbabwe has also embarked on a number of efficiency improvement programmes, some of which have been funded by the Bank. To some extent, therefore, it can be concluded that project implementation is consistent with policy. However, MDB policy has failed to address the issues of technology transfer and energy efficiency in the transport sector.

Often, multilateral institutions such as the World Bank are concerned with large power sector projects, which ultimately benefit the commercial and urban domestic sectors. The Bank has conducted studies on the rural sector but concrete actions have not materialized from such studies. There is a certain amount of interest now in renewable energy, and a number of players have latched onto this bandwagon. But it is questionable how much will be achieved. Donors have often taken a very narrow view in promoting renewables by merely presenting the hardware. What is required is a comprehensive approach to technology commercialization, one that encompasses research, development, demonstrations and market diffusion, and this can require almost a decade to complete. Too often technologies have been promoted before they are proved to be effective, and very little attention is placed on developing indigenous institutional capacity to commercialize and deploy them. The Bank and other donors therefore need to reassess their involvement in the development of energy for rural areas.

The private sector also needs to get further involved in the development of renewable energy. Current interventions by the private sector are limited to the installation and manufacturing of a limited range of PV components. Private sector companies could begin sponsoring awareness-raising efforts by NGOs and also providing training for rural people to maintain their systems. While it may be argued that such a position would undermine the profitability of the private sector, such a view would be short-sighted, as the private sector can only gain from well-maintained systems, which tend to advertise themselves. Such a move would also be good public relations as currently some firms are failing to provide adequate follow-up of the systems they have installed.

It is clear from the above discussion of the institutions involved in the development, sourcing and delivery of energy that many constraints exist, but that close cooperation, coordination and implementation of decisions is imperative if success is to be achieved in the development of this important sector. The need is especially more apparent in the rural areas, where energy issues have never been addressed in a coherent way. Interventions in this area have at best been scanty, and lacking in real focus. While the Department of Energy has taken some steps in addressing the issues through its various initiatives, such as the biomass energy strategy formulation process and the various energy technology projects, there is still a lot that needs to be done. Another problem is the weakness of links between energy planning and macro-economic planning. This has largely been due to a lack of a strong institution in government to take the lead in preparing and implementing decisions on investment priorities, regulations, technology transfer, and skills development in the energy sector. While the National Economic Planning Commission is mandated to provide macro-economy-wide planning, its orientation is towards the traditional growth sectors such as agriculture, mining and manufacturing, so that energy is not always similarly prioritized.

For the energy sector in Zimbabwe to address the issue of energy development adequately, a number of issues need to be addressed. First, there is a need for the reinforcement and streamlining of all energy institutions at sector and sub-sector levels and for the strengthening of professional competence. Institutional relationships also need to be rationalized to facilitate the exploitation of opportunities in the energy sector. A coordinated approach to the problem will ensure that resources are used efficiently. The Ministry of Transport and Energy also needs to develop consistent energy sector policies out of which a consistent energy planning methodology will flow. Current efforts by the Department of Energy in this direction are indeed welcome, as they are long overdue. The Department of Energy's capacity to coordinate all energy sub-sector plans and integrate these with overall national economic development objectives should be addressed. The department's capacity in project evaluation, monitoring and implementation, and sector regulation should also be addressed at the same time, as, without this, coordination will be very difficult. The department's capacity is, however, constrained by the lack of resources. The only avenue for financing is the Public Sector Investment Programme and sporadic donor efforts. This low level of funding has meant that salaries have been kept low, which in turn has resulted in an exodus of qualified people from the department at a time when the department needed to increase its capacity.

People's participation and consultation in the energy sector needs to be facilitated through the development of local level institutions, such as Ward Development Committees (WADCO), Village Development Committees (VIDCO), church groups and women's groups, which can link up with national ones for local level energy development. The NGO sector needs to be recognized as an equal partner in the development of the energy sector, and strategies should be put in place by both government and the NGOs to address the problem. Finance plays an important role in the development of the energy sector. Substantial financial resources are often required to support energy development, whether large-scale or decentralized power generation. In the commercial energy sector, such as electricity, the raising of finance is not often a problem, as the multilateral development banks' portfolio adequately covers this sub-sector. The problem arises with decentralized alternative energy technologies. Funding for these activities has in recent years declined, although there is now a flurry of new initiatives to fund alternative energy development. The Global Environment Facility (GEF) is emerging as the heavyweight in this category, having so far financed numerous alternative energy projects throughout the world in a bid to address climate change.

Decentralization of planning structures is an important initiative that needs to be intensified. The system of planning at VIDCO and WADCO level is an important entry point into decentralization, but needs to be strengthened through building the capacity of the people at that level to understand and to plan effectively for the issues that confront them.

There also needs to be reform related to legislation affecting the functioning of the energy sector. Such reform could include making it easier for independent players to participate in power generation, to become manufacturers of various components, and to make the use of renewables more attractive than conventional forms of energy.

It is imperative that institutional reforms be carried out in order for Zimbabwe to meet its current challenges, but consensus needs to be established on how this should be carried out. In conclusion, based on the fact that the banks' energy sector policies have rather limited environmental protection biases and that Zimbabwe's policy is more concerned with the local environment, it can be safely concluded that project implementation has been in line with policy requirements. On the financial side the Bank tried as much as possible to adhere strictly to its policies. From the social and environmental point of view, however, the Bank has not been consistent. The Bank talks a lot about poverty alleviation and gender-sensitive projects but the investments in Zimbabwe so far do not show any of these biases. In addition, in most cases, Bank-funded projects have not been at all responsive to sustainable energy issues, such as climate change. It is apparent that most projects have been funded on the basis of their ability to generate a financial return for the Bank, and this explains the general trend in funding towards large-scale generation, distribution and transmission projects.

Note

1. Analysis for Zimbabwe prepared by Bhekimusa Maboyi, ZERO, PO Box 5338, Harare, Zimbabwe.

6

China[1]

Opportunities for China's Power Sector

To fuel an annual gross domestic product (GDP) growth of 9 per cent during the past two decades, China has dramatically increased its energy use and expanded its domestic production and imports of energy. Now, China ranks as the second-largest energy consumer and producer in the world, following the USA. By 1997, China's primary energy consumption reached 37.1 quads, next only to 94.0 quads consumed in the USA (US Department of Energy 1998). At this time, China's energy sector is self-sufficient. Exports and imports of energy products are small compared to overall consumption, although the amount of imported oil and gas products and exported coal are expected to rise. China's self-sufficiency strategy has led it to depend heavily on coal for national energy needs. In 1996, coal accounted for 72.8 per cent of total consumption, followed by oil with a share of 19.9 per cent, and hydropower contributing 5.1 per cent. Natural gas and nuclear power contributed only 2.1 per cent and 0.4 per cent respectively (Chandler et al. 1998). The country's industrial sector is the main energy consumer, using about 60 per cent of end–use energy. The transportation sector accounts for 12 per cent of end–use energy. In contrast to most Western countries, China's residential and commercial sectors consume only about 23 per cent of end–use commercial energy, and the agricultural sector consumes only 5 per cent, despite the fact that much of China's population continues to be involved in food and animal production (State Economic and Trade Commission, China 1996). National economic reforms, which seek to restructure China's industry to higher added value enterprises, are expected to reduce the share of energy consumed by the industrial sector. On the other hand, the shares of energy consumed by the transportation, residential and commercial sectors are expected to rise due to increasing incomes. Commercial energy consumption for agriculture is projected to be stable over the next decade.

The Chinese government has declared its intention to maintain high economic growth into the next century. To sustain this growth, the energy sector is planned for expansion. Commercial energy supply is expected to reach about 44.3 quadrillion Btu (quads) in the year 2000, and 72.5 quads by 2015, an increase of 4.7 per cent annually to 2015 (US Department of Energy 1997). Coal is expected to retain its importance in China's energy production mix, with its share rising to 77.4 per cent of the total by 2015. The hydropower share is expected to reach 6.2 per cent, and natural gas is projected to grow to about 4.1 per cent by 2015, as China begins to take greater advantage of its large domestic reserves. China has an ambitious plan to develop its nuclear energy, with the share from this risky technology officially expected to reach 1.6 per cent by 2015. As domestic petroleum reserves shrink, petroleum's share will fall to 10.7 per cent (ibid.).

China's power sector has grown in step with the national economy. For the last ten years, about 10 GW of generation capacity has been added annually. By 1998, the country was adding 15 GW per year. As a result, China's total generating capacity has reached 250 GW, with power generation amounting to 1,335 terawatt hours in 1998 (*People's Daily*, 12 December 1998). Per capita electricity consumption increased at an annual rate of about 9 per cent over the past decade. Although rapid growth has been the hallmark of China's electricity programme, per capita electricity use still remains low, roughly one-third of the world average. Further, there are more than 60 million people, mainly in rural areas, who still do not have access to electricity (Xinhua News Agency, 8 December 1998). According to a recent international study (Chandler 1998), China's power demand is expected to reach 1,390 terawatt hours in 2000, 2,500 terawatt hours in 2010, and 3,210 terawatt hours in 2015. Compared with earlier forecasts conducted by China's energy research organizations, this forecast considers recent changes in China's macro-economic development, which build in higher electricity demand. While these projections are higher than official figures, it is likely that they more accurately portray the future, unless policy interventions to promote greater energy efficiency are made. In order to meet the country's growing electricity demand, an addition of nearly 380 GW of capacity between 1999 and 2015 is needed. This expansion, however, requires a huge amount of investment. One report estimates that during 2000–10, China's power sector will need almost $315 billion to meet the capital requirements of its capacity expansion plans (Chandler et al. 1998). China assumes that 20 per cent of its power sector capital requirement can be provided by foreign funds. This amounts to an annual flow of $4.2 billion investment funds from foreign sources between 2000 and 2015. This would be an exceptionally high capital commitment, both in domestic and international terms, for an electricity sector's growth.

While coal will continue to dominate the energy sector, China's energy supply options will become diversified, with increased exploration for and imports of oil and natural gas. Also, large-scale hydroelectricity and nuclear technology will be promoted by the government in order for the country to reach its electricity expansion goal. In addition, China's energy sector, especially the power sector, will continue its transition from a centrally planned system to a market-oriented one. Reform of the country's electricity tariffs and its foreign exchange system and the power project approval procedure and regulatory framework will be accelerated as well. Currently, privatization is being discussed for power generation.

China has been a member of the World Bank since 1981, and has been the Bank's largest borrower of investment financing since 1992. In FY98, World Bank Group (IBRD and IDA) lending to China totalled US$2.62 billion (US$2.32 billion in IBRD and US$293.4 million in IDA), bringing cumulative lending as of 30 June 1998 to about US$31.13 billion, of which US$21.6 billion is IBRD and US$9.53 billion is IDA. Among the US$2.32 billion IBRD loans to China in FY 1998, about one-quarter (US$550 million) directly went to energy development (World Bank 1999). In the past few years, lending for environmental protection has become the fastest-growing area of the World Bank's programme in China. In FY98, four projects with lending totalling US$350 million were approved, benefiting the urban environment, energy conservation, and coastal zone resources. The Bank's most recent assistance to Chinese policy-makers is its environmental study *Clear Water, Blue Skies*, produced in close collaboration with China's State Environmental Protection Administration and the State Development Planning Commission. Other studies supported by the World Bank include: the environmental impact of coal use, greenhouse gas emissions control, and biodiversity conservation. Based on the findings of these studies, the World Bank has earmarked US$3.4 billion to improve air and water quality in China's cities. Furthermore, Bank-assisted power projects now routinely incorporate environmental standards in their design. The Asian Development Bank (ADB) has played a similar role to that of the World Bank in China. China's energy sector is also the principal recipient of ADB loan support. Currently, six power projects are under way with ADB financing. These include two fossil fuel plants, two transmission projects, a wind farm and an environmental control project.

Social and Environmental Challenges

As China's economy expands to meet the growing needs of its population, the country faces great challenges in balancing its goal of economic growth with energy, environmental and social sustainability. Several issues will need to be addressed for sustainable development to take place. Financially, expanding energy supply and distribution systems at the planned rates by national authorities will place significant investment burdens on the current and next generation. China's intention to increase power capacity by 380 GW between 2000 and 2015 at a cost of more than $300 billion is to be juxtaposed with the fact that available funding from multilateral organizations has been below $1 billion per year in the 1990s. Further, there is the problem that well-developed domestic capital markets do not exist yet to finance energy projects of this magnitude. And foreign participation in China's energy projects will be hampered by the lack of a tested regulatory and legal framework that could address the risks associated with the foreign investment. Along with sizeable financial constraints, there are also major threats to environmental sustainability. Many of China's environmental problems stem from the increasing and inefficient use of fossil fuels, especially coal. For years, the people of China, particularly those living in cities, have breathed air with harmful levels of sulphur dioxide and particulates that are two to five times World Health Organization standards. In addition, intensive and inefficient use of coal resources is creating one of the world's most serious acid problems. It is estimated that acid rain now affects nearly 40 per cent of China's land area and causes over $13 billion of damage annually to the country's forests, farms and human health (Chandler et al. 1998). In producing approximately 14 per cent of global CO_2 emissions, China is now recognized as the world's second largest emitter of greenhouse gases, trailing behind only the USA. It is likely that China will soon become the world's leading source of CO_2 emissions if current trends of energy use continue.

China's power sector is also poorly coordinated within the government. Responsibilities for power-related energy development are widely dispersed among various agencies, making the development of a systematic and comprehensive energy strategy very difficult. If China is to meet its energy needs for sustainable development in the future, it needs to create a transparent organizational foundation. In addition to the financial, environmental and institutional challenges facing China's power sector, the country is confronting great difficulties in providing a reliable electricity service to its rural areas, where three-quarters of the population live. Although the government of China has made significant efforts to bring electricity to its rural residents, 40 per cent of the country's rural households (90 per cent in the country's western provinces) still do not have access to power grids because of the prohibitive costs of extending electricity services to remote rural areas. Overall, these issues – investment gaps, environmental threats, institutional barriers and distribution problems – have surfaced to challenge China's energy sector. Addressing these issues, however, cannot rely upon the fossil fuel-based, centralized options. Pursuing energy, social and environmental sustainability while continuing economic development in

China requires increased use of alternatives such as energy efficiency and renewable energy.

Sustainable Energy Options

While current trends towards nuclear and fossil power may only add to the environmental and social problems highlighted in the previous section, the emerging interest in developing renewable energy and energy efficiency options by China's government offers a more sustainable approach to energy issues. Energy efficiency and renewable energy options are emphasized in new development plans such as *China's Agenda 21*, *Guidelines of the Ninth Five-Year Plan*, and the *Long-Term Objectives for Economic and Social Development of China*. Regulatory and market-based incentive policies can be expected to encourage the development and deployment of energy efficiency and renewable energy. The implications of these conflicting trends need to be recognized. Developing large-scale hydroelectric and nuclear projects is not only economically risky but also environmentally problematic. A study of China's power sector released recently by the US Pacific Northwest National Laboratory indicates that the capital cost of building large-scale nuclear power or hydroelectricity could be 45 per cent higher than that of natural gas combined-cycle units. The cost could be much higher if the environmental and social impacts of developing large-scale nuclear and hydro programmes are included (Chandler et al. 1998).

Energy efficiency and renewable energy could be the least-cost options for China to address its growing energy needs. The country has a great potential in improving energy efficiency and developing renewable energy options. Despite its impressive success in adopting efficiency measures, China still has one of the highest energy intensities in the world (the country on average requires three or four times as much energy input per unit of output as the developed countries) (*China's Agenda 21* 1994). This fact indicates that China has great opportunities to improve its energy efficiency. In addition, it has abundant renewable energy resources. The country's geothermal reserves are equivalent to 3 billion tons of coal equivalent, but only 0.01 per cent of this resource is being tapped. China's total windpower potential is estimated at 1,600 GW. This is over eight times current Chinese electricity generation capacity. Similarly, the prospect for photovoltaic (PV) technology in China is strong. Most parts of China receive quite high levels of solar radiation, averaging 1,668 kWh per square metre annually (ibid.). Developing China's abundant renewable energy resources would provide a sustainable approach to addressing the growing energy needs of the population, especially those living in rural areas.

Although China has made efforts in examining alternative energy options, it has focused its efforts primarily on efficiency improvement of its industrial sector from a supply perspective. In addition, its renewable energy efforts have concentrated on the development of large-scale, grid-connected renewable energy options. Research conducted by the Center for Energy and Environmental Policy at the University of Delaware shows that China would gain more social, economic, energy and environmental benefits by implementing a full range of efficiency measures in its industrial, transportation, building, residential and commercial sectors and by increasingly developing decentralized, off-grid renewable energy technologies, especially in the country's rural areas (Byrne et al. 1996).

References

Byrne, John, Bo Shen and Xiuguo Li (1996) 'The challenge of sustainability: balancing China's energy, economic and environmental goals', *Energy Policy* 24 (5): 455–62.

Chandler, William, Zhou Dadi, Jeffrey Logan, Guo Yuan and Shi Yingyi (1998) 'China's electric power options: an analysis of economic and environmental costs' (draft final prepared for the W. Alton Jones Foundation), Washington, DC: Battelle Memorial Institute and Beijing Energy Efficiency Centre, Energy Research Institute of China, June.

China's Agenda 21: White Paper on China's Population, Environment, and Development in the 21st Century (1994), Beijing: China Environmental Science Press.

People's Daily (overseas edition), 12 December 1998.

State Economic and Trade Commission, China (1996) *China Annual Energy Review 1996*, Beijing.

US Department of Energy (1997) *Country Energy Profile: China*, Washington, DC: Energy Information Administration, October.

—— *International Energy Outlook 1998*, Washington, DC: Energy Information Administration.

World Bank (1999) 'China and the World Bank' available URL: http://www.worldbank.org

Xinhua News Agency, 'China's stance on global climate change', 8 December 1998.

Notes

1. It was not possible to find a Chinese NGO to write freely and openly about the energy and development issues. This chapter was prepared by Dr Bo Shen, Center for Energy and Environmental Policy, University of Delaware, Newark, DE 19716, USA.

7

India[1]

The Indian Power Sector in Crisis

Most of India's power generation and distribution is in state hands. Integrated utilities, owned by the State Electricity Boards (SEBs), account for 95 per cent of power distribution and 62 per cent of power generation in the country. Federal companies account for 30 per cent of generation and inter-state transmission, with 8 per cent generated and distributed by private utilities. Together these utilities constitute an installed capacity of 90,000 MW, supplying over 82 million people in thousands of cities and towns as well as 500,000 villages. Consumption of electricity is growing at around 7 per cent per year. Over the last five decades, the Indian power sector successfully doubled its infrastructure capacity every nine years to meet the growing demand. But now, a combination of factors, including the system of government subsidies and tariffs that is in place, theft, inefficient energy use and the rising growth in demand amount to a severe financial crisis in the power sector. High commercial losses are mainly attributable to a heavily subsidized agricultural tariff, which is based on the measure of connected load rather than on actual energy consumption. This can lead to poor end-use efficiency and also allows power to be stolen or power consumption to be 'hidden' from power utilities. Paradoxically, a large portion of the agricultural subsidies, which are defended on the grounds of equity, is awarded to wealthy farmers and, so far, attempts at reform have met with little success.

Responding to this financial crisis, the government has taken steps to attract private sector investments. First, in 1991, it attempted to increase generation capacity by opening the sector to independent private power producers (IPPs). The initial response from the private sector was very encouraging, but measurable progress, in terms of increased capacity, remains very limited. This is primarily attributable to the fact that the SEBs, which were expected to be the buyers of IPP power, were in such a poor financial condition at the time of restructuring. It is evident now that unless the problem of the financial viability of the utilities is tackled, private capital will not materialize. In keeping with the World Bank's position, governments are moving towards a radically altered structure for the power sector. The three-pronged restructuring strategy firstly involves 'unbundling', that is, the separation of the generation, transmission and distribution functions of state utilities. Second is the establishment of independent regulation, and privatization of generation, transmission and distribution units. Many state governments, using MDB loans, are planning to adopt the same restructuring model and the central government has created an enabling environment for such reforms by enacting the 1998 Electricity Regulatory Commissions Act. Now the central government is suggesting that all states work out SEB reform plans in a time-bound manner. In a nutshell, installing an independent Regulatory Commission and implementing privatization are the priorities, while environmental and social issues have not received due attention. Such issues acquire import-ance only when opposition arises against specific projects, but even then the government tends to push the project to completion at all costs, ignoring the valid concerns of stakeholders. In the absence of any mechan-isms for public participation in planning and policy-making, this system of priorities has become entrenched.

Renewable energy (RE) and energy efficiency (EE) measures have not found a place in mainstream energy planning in India. The current efforts of governments are restricted to conventional mechanisms such as providing concessional finance and tax subsidies, and awareness-creation measures. Further, these are focused on the conventional RE and EE technologies that have already been commercialized. Governments at the federal and state levels have established specialized agencies to promote RE devices and schemes. Until recently, these agencies have focused on such technologies as solar cookers, PV lanterns, PV street-lights, solar-water heaters, small diesel generator (DG) sets based on wood-gasifiers, megawatt-range biomass-based power generation, public education and disbursing subsidies. The Indian Renewable Energy Development Agency (IREDA) is a node of the government of India that provides concessional finance for RE techno-logies and schemes. MDBs are providing a line of credit of around $300 million through IREDA to develop mini- or micro-hydro plants, wind farms, PV installations and co-generation projects. In addition, the Global Environment Facility (GEF) is supporting a 35- to 40-MW solar thermal project for power generation. The present installed capacity of the RE-based power projects in India is around 1,100 MW, which is expected to grow to 3,000 MW by 2002. Of the present capacity, wind contributes 825 MW and small hydro 135 MW, with the remainder generated from biomass and solar energy sources. Although these are welcome developments, there are some crucial shortcomings. For example, due to the tax and other concessions, wind projects have become highly profitable, with an internal rate of return over 50 per cent and incentives for cost padding. Recent sales tax concessions offered by several state governments amount to government subsidies for half the capital costs of RE projects. Such high expenditure has not borne the commensurate benefits and could have been invested in more effective RE activities and measures that have multiplier effects and long-term benefits, such as research and development, demonstration projects, simplifying enabling procedures and institutions for projects and training.[2]

The Industrial Development Bank of India (IDBI) has run a programme since the mid-1980s that provides concessional loans for industrial EE projects, which the ADB has supported by way of a line of credit. The government has made it mandatory for companies to declare in their annual reports their energy intensity (energy consumption per unit of production) and steps taken for improving efficiency. The Energy Management Centre (EMC), was set up under the Ministry of Power to promote energy efficiency. It has largely been a promotion agency advertising and facilitating the organization of seminars, workshops, and some research. Power-sector planners have taken steps to improve the energy efficiency of existing power plants and installed capacitors in the transmission and distribution networks. While these efforts have met with some success, the problem remains that such projects are not incorporated into national plans to increase energy capacity, but remain peripheral. Demand-side efficiency, for instance, has largely been a non-starter in India. The EMC, independent researchers and World Bank consultants have carried out several DSM studies, but few schemes have reached the implementation stage, with the result that SEB confidence and willingness to commit resources remains low. As part of the wider reform process, the World Bank has extended a loan of $100 million for DSM projects to the state of Orissa, half of which will be spent on metering! Similarly, many other state reform projects funded by the MDBs have a DSM component but the failure of the government to set standards for energy efficient appliances remains a problem. Consumers, eager to conserve energy, were misled by electronic ballasts with low life and poor performance and lost confidence in investing in energy saving technologies in general.

The First and One of the Largest Borrowers

India is the original signatory to the Bretton Woods Agreements in 1944, under which the World Bank and the IMF were established. India is one of the largest borrowers from the World Bank, in terms of both IBRD loans and IDA credits. Between 1949 and 1997, the Bank has extended 179 loans and 257 IDA credits to India totalling approximately $18.9 billion and $22.5 billion respectively. The Asian Development Bank began lending to India, now a major borrowing country, in 1986. Like the World Bank, the ADB has two lending instruments: Ordinary Capital Resources (OCR), a standard loan instrument, and the Asian Development Fund (ADF) which provides concessional loans. Since 1986, through 42 public sector loans for 37 projects, the ADB has provided financial assistance to the tune of $6.96 billion to India.[3] The energy sector (oil, gas and power) accounts for the largest share, 48 per cent of IBRD loans to India, but only 11 per cent in the case of IDA. Of these, the power sector accounts for three-quarters of World Bank support for the energy sector in India.[4] The National Thermal Power Corporation (NTPC), a central government-owned thermal generation utility formed in the mid-1970s, is the single largest borrower of World Bank funds. During the mid-1980s, the World Bank moved away from project loans towards SEB loans and then, around 1990, shifted to sectoral loans. After 1994, the Bank decided not to support the state-level power sector unless the state agreed to adopt its reforms model. The following is a category-wise but brief discussion of the WB's loans in the energy sector.

SEB loans This class contains loans given to utilities owned by state governments such as Uttar Pradesh SEB, Kerala SEB and Maharashtra SEB. These loans were typically given in the mid-1980s and ranged between $200 million and $500 million, supporting projects requiring large investments such as power generation or transmission. These loans were linked to the achievement of certain performance standards in terms of improving operational/financial performance and commercial viability (by measures such as improved financial discipline, and tariff rationalization). The Bank had to cancel five out of six recent loans, however, due to non-compliance with loan conditions by the SEBs/state governments.

Thermal power loans These loans, mostly to the NTPC, were extended for increasing coal-based thermal power generation from large pit-head projects. Social activists and organizations representing project-affected peoples have seriously criticized the WB loans to NTPC for power plants in the Singrauli region. The NTPC has developed more than 4,000 MW of power generation capacity in the region. This, along with the coal-mining activities, has led to severe environmental and rehabilitation problems.

Hydropower loans The World Bank has supported hydropower development in India since the early 1960s through nine projects, involving loans and credits totalling $1.37 billion (net of cancellations). The WB has faced severe criticism for its involvement in these hydro projects, mainly in relation to rehabilitation and environmental issues. Opposition and struggle by project-affected people in the Sardar Sarovar Project in the Narmada valley was a watershed in the history of the Bank. The struggle drew worldwide attention to the problems of project-affected people, to fundamental issues in rehabilitation policy, to environmental and social impacts of large dams, and to the Bank's failure to adhere to its own policies. The Bank was forced to appoint an independent review committee (popularly known as the Morse Committee) to review the project and rehabilitation situation. Following the serious social and ecological damage associated with the project, pointed out by the Morse Committee, the Bank eventually withdrew its support.[5]

Transmission loans Ten projects under this category, amounting to $2 billion in loans and credits, have supported the much-needed investments for power transmission. Over the years, Indian utilities have grossly under-invested in transmission. These loans provided mainly for addition of inter-state transmission links and links for evacuating power from large power projects. These projects also had components for improving the dispatch and coordination functions of the regional load dispatch centres and for ensuring commercial discipline (in the case of the Powergrid System Development Project).

Coal sector loans Projects under this category provided loans to private and government mining companies in the early 1960s and mid-1980s to enhance mining capacity by adopting modern equipment and techniques. The Coal India Environmental and Social Mitigation Project, approved in 1996, needs a special mention. This $63 million IDA credit is exclusively aimed at mitigating the social and environmental impacts of 25 coal-mines (including mines in the Singrauli region). Under this project, plans will be prepared for environmental regeneration, for rehabilitation of affected people, and for economic development for indigenous people in the area. In addition, a special fund to finance mitigation and to remedy the environmental and social impacts will be established. The capability of Coal India Limited to handle such issues is also expected to be upgraded through technical assistance.

Apart from the main project categories mentioned above, the Bank has also funded projects in the areas of rural electrification, energy efficiency and renewable energy. The three rural electrification projects (totalling about $0.5 billion) supported expanding and strengthening rural distribution to increase access and to improve the quality of power supply in rural areas. The loan for the India Alternate Energy Project provided $216 million (including $26 million GEF grant) for co-generation, small hydro, photovoltaic (PV), and wind-farm development in the private sector. This was a credit line for IREDA (the government agency for funding renewable projects). An International Development Association (IDA, the World Bank's soft-loan window) credit line to IREDA, amounting to $70 million, for the India Renewable Energy Resources Development Project supported private initiatives in renewable sources. Mini- and micro-hydro projects of 116 MW are expected to be commissioned through this project.

Recent World Bank Projects

Coal sector rehabilitation This loan of $535 million to Coal India Limited, a central government company producing 90 per cent of India's coal, will support the purchase of equipment for over 20 coal-mines to increase production. The removal of import duty for coal and the increased role of the private sector in the mining industry are among the prominent loan conditions. This project is supported by a Coal India Environmental and Social Mitigation Project (CESMP) loan, involving $65 million, as mentioned above. Successful implementation of CESMP has been laid down as a condition for approving this loan. Discussions with NGOs have also been made mandatory by the executive directors of the Bank to initiate the loan negotiations.

SEB restructuring Under the power sector reform agenda, the World Bank is providing large loans to state utilities in the states of Orissa, Andhra Pradesh and Haryana, ranging from $300 million to $1 billion each. Through these loans, the Bank is pursuing radical reforms in the institutional structure, tariff policy and ownership of utilities. The existing state-owned power utilities will be broken up into separate generation, transmission and distribution companies and gradually privatized. An independent Regulatory Commission (RC) will also be set up, with wide-ranging powers including those of setting bulk and retail tariffs and ensuring supply- and demand-side efficiency. The main responsibility of the RC will be to ensure the smooth commercial operation of utilities and to adjust tariffs so as to generate sufficient internal resources for expansion, while protecting consumer interests. The government will have to reimburse the utilities if it wants to offer subsidies over and above the cross-subsidy permitted by the RC. The restructuring process is nearing completion in Orissa, where licences for generation and distribution have been issued to unbundled utilities, generation companies have been privatized, and bids for privatizing the distribution companies have been awarded.

PowerGrid II Under this loan of $400 million, the Bank would support the PowerGrid Corporation of India, a government-owned central transmission utility for inter-state transmission. This would partly finance the investment needed – about $4.5 billion over the years 1997–2002 – for, among other things, strengthening existing inter-regional links as well as for establishing new links and improving control systems. It is expected that the PowerGrid will fully establish 'loose power pools' and allow IPPs open access to the national grid to sell power to any state in the country. Strengthening the national grid will also allow 'merit-order dispatching' of plants and alleviate power shortages by allowing power transfer between different regions.

Renewable Energy II This is the sequel to the first RE loan and has a similar structure. Under RE II, the Bank would provide $170 million to public and private investors to develop approximately 200 MW of capacity through mini- and micro-hydro projects. A small part of the loan would also be made available for DSM projects. This loan will be used by IREDA to sub-lend at near-market level interest rates of 16.5 to 17 per cent.

India solar-thermal project Under this project, the Bank, supplemented by funds from the Global Environment Facility (GEF), is supporting a large-scale solar-thermal project in the desert state of Rajasthan (along with a fossil fuel plant). A 35- to 40-MW solar-thermal plant (with parabolic trough 'Luz' technology) and 100-MW combined cycle gas turbine (CCGT) plant (using fossil fuels) will be established. These plants will be operated at 80 per cent plant load factor (PLF) by a private party on commercial principals. The loan includes a GEF grant of $49 million and some technical assistance. The Indian government will provide an additional $20 million subsidy to compensate the IPP for the incremental cost of the solar-thermal project.

The Asian Development Bank in India

Asian Development Bank (ADB) lending operations in India began in 1986, and lending for the energy sector for 15 projects is about $2.6 billion, or 40 per cent of ADB lending to India. Until the early 1990s, the ADB mainly supported thermal power projects and since then it has concentrated on the hydrocarbon sector, with projects to develop gas-fields and reduce gas flaring. Currently, the hydrocarbon portfolio, at $887 million, is the largest beneficiary of ADB support to the Indian energy sector. Apart from the hydrocarbon and thermal power projects, the ADB has also supported 'energy efficiency improvement projects' in the power sector and industries. These include the ADB credit line of $150 million to the Industrial Development Bank of India (IDBI) for financing a complete change of industrial process towards energy efficiency. This loan will also strengthen evaluation procedures used by the IDBI for energy efficiency projects through the help of consultants. Other loans in this category include the Power Efficiency Loan of $250 million to improve supply-side efficiency. As described in the Country Assistance Plan for 1999 to 2001, the ADB plans to maintain total lending to India at the levels of $900 to $1,000 million a year for all sectors combined (ADB 1999). The ADB will directly support restructuring of the state power sectors in Gujarat and Madhya Pradesh (MP). Through these projects, issues such as irrational tariffs and operational inefficiency are expected to be addressed by instilling competition, commercialization and private sector participation. The ADB reform agenda is quite similar to that of the World Bank, with a $250 million loan to the Power Finance Corporation (PFC) for supporting the state-level restructuring projects in other states. Apart from these re-structuring projects, it will support combined cycle power generation (loan of $250 million) and an industrial energy efficiency project (loan of $200 million).

The 'Energy Shortage Psychosis'

The World Bank and other MDBs have been helping to perpetuate the so-called 'energy shortage psychosis', which precipitates *ad-hoc* decisions to implement large, centralized, supply-side energy projects based on conventional fuels. This tendency towards large projects is reinforced by the absence of integrated resource planning and non-market measures to promote the efficiency and renewable energy options. It is primarily from this imbalance that the environmental and social implications stem. In response to these criticisms, the MDBs have taken up some programmes on renewable energy and efficiency. However, these programmes remain peripheral to mainstream policy and so cannot address the energy needs of such large rural populations with high levels of poverty. The reform agenda presents problems in terms of the content of the reforms and the manner in which they are implemented, which has been characterized by a lack of participatory debate and open-mindedness on the part of MDBs. The outreach activity conducted jointly by the MDBs and the implementing governments could best be described as exercises in promoting pre-determined outcomes. Although there are successful models of public participation in policy-making and planning exercises in such places as California and Kerala, MDBs have not made them standard practice.[6] In terms of the content of the reform packages, the privatization model is expected to eliminate the ills of inefficiency, indiscipline, theft and corruption. However, in practice, the success of the MDB prescription is heavily dependent on the power of the authorities to apply honest and capable regulation without being subject to counter-checks and monitoring. The vivid contradiction between, on the one hand, the aggressive stand taken by the MDBs on privatization and restructuring and, on the other hand, the absence of support for efficiency or renewable energy is too blatant to miss. These contrasting attitudes could be described as 'just do it' (for privatization) and 'just think it' (for renewable energy and efficiency).

The tendency of the MDBs in response to criticism to effect peripheral policy changes and minor actions is not commensurate with their privileged position and responsibilities. In order to be true to their developmental goal of poverty reduction and to inspire confidence in their commitment to democracy, MDBs should undertake three crucial tasks. First, they need to get involved in a truly participatory dialogue with people and grassroots-level functionaries. It is heartening to see the recent acknowledgement that the understanding of the ground realities and peoples' aspirations is not readily available in the Ivy League universities. Neither can this knowledge be imported through a few consultations with NGOs and ground-level functionaries. It could emerge only through a participatory dialogue between the analytical skills and theoretical capabilities of experts within the MDBs, and the experiential insights of ground-level functionaries and people. This dialogue must be conducted with mutual respect and on equal grounds to help analysts and professionals in the MDBs transcend the limitations imposed on them by their training and location (social location, class location and geographic location). It will also help them to view issues of development and livelihoods, as well as life in general, from the standpoint of the poor and the disadvantaged. This brings us to the second critical responsibility of MDBs, to examine and, wherever necessary, to restructure the very foundations of their vision, policies and programmes. In other words, the second responsibility requires that MDBs fundamentally restructure their rhetoric, policies and actions to pursue the secure and sustainable livelihoods of people in low-income communities. Third, they must translate this policy transformation into practice by trial and research in collaboration with affected communities to transform the 'desirable' into the 'viable'.

Notes

1. This chapter was prepared by Shantanu Dixit, Girish Sant and Subodh Wagle, Prayas, Amrita Clinic, Athawale Corner, Karve Road, Corner Deccan Gymkhana, Pune – 411 004, India.

2. One such mechanism, which was adopted in the UK, compels utilities to buy a certain percentage of their generation from non-fossil sources. By adopting such measures, additional cost for non-fossil generation could be supplemented through subsidies.

3. Asian Development Bank, *Country Assistance Plan – India, 1999–2001*, Manila, 1999.

4. The power sector is the major user of coal and consumes more than 60 per cent of coal production in India. Thus, the power sector also benefits by the World Bank's loans to the coal sector.

5. Officially, the government of India requested that the loan be cancelled.

6. In the state of Kerala, responding to popular demand, the government conducted a state-wide bottom-up programme to evolve development plans of villages in which larger public participation was ensured.

8

Indonesia[1]

Indonesia's Power Sector

Almost everything that takes place in the electricity business in Indonesia is the product of PLN, Indonesia's state-owned electricity company; from power generation and transmission of 18,800 megawatts of electricity to distribution to 22 million consumers in an area nearing 2 million square kilometres. As an agent of the government, PLN implements social policies such as the electrification of remote rural areas, which is not profitable, while also working on more profitable ventures with the industrial sector. Outside of the main Java–Bali power system, infrastructure is less developed with relatively limited networks and numerous scattered small-capacity power generators, which incur higher operation costs. PLN employs around 58,500 people. Outside of PLN, the government role in the power sector includes the provision of loans via international agencies, and setting electricity policy and regulations. A problematic feature of the government's relationship with PLN is that on the one hand, as its owner, there is an expectation that government policies should be beneficial to PLN, while on the other hand, the government is obliged to prioritize the needs of consumers. These are not always mutually exclusive priorities but can result in a conflict of interest, exacerbated by an inefficient and untransparent bureaucracy. The PLN profited by more than one billion Indonesian rupiahs annually before the monetary crisis hit Indonesia and cost PLN nearly six billion rupiahs in 1998 alone. PLN was forced to seek extra subsidies from the government in order to keep functioning and pay liabilities to foreign credit sources, which became heavier when projects contracted out to private companies were marked up. Private companies were mostly chosen without competitive or transparent biddings and project costs were exaggerated in favour of companies with poor financial performance. In a further effort to solve its financial woes, PLN then proposed to increase the electricity tariff and fuel price. The government had designed the electricity tariff structure carefully, allowing a cross-subsidy so that small consumers would not bear the brunt, but this did not prevent the outcry that followed from a public that had already been hit hard by the economic crisis. The government was forced to revise the plan and lower the increases and PLN is now under the public spotlight, facing accusations of inefficiency, collusion, corruption and nepotism, particularly in relation with power purchase agreements (PPAs) between PLN and independent power producers (IPPs). The public demanded that PLN begin to supply reliable electricity, become transparent and efficient in its dealings. A survey by the World Bank found that the average PLN industrial consumer must endure frequent outages totalling between 40 to 90 hours per year. Added to this are distortions in voltage, frequency and waveform, as well as lengthy and uncertain waiting times for new connections. These have caused high quantities of captive power being used by companies to back up PLN supply. Approximately 6,000 MW of captive generation has been acquired. The relatively cheap diesel fuel price has also encouraged the utilization of captive generators, which are mostly diesel fuelled. Unfortunately, these small-scale generators are inefficient and contribute to air pollution.

Subsidies have always been a controversial issue in Indonesia. Needed by many low-income families for vital goods like food and energy, they often miss their target and generate new problems. The electrification rate in Indonesia is approximately 50 per cent, with most consumers living in urban areas. Eighty per cent live within the Java–Bali network and are thus the beneficiaries of the electricity subsidy. Low-income families outside the grid miss out while urban dwellers are subsidized whether they need to be or not. There is a good case for reallocating this subsidy to more needy areas. Considering that a large portion of PLN's expenditure is for paying IPPs, this could be regarded as a subsidy for large private companies.

Private participation in power sector in Indonesia began in 1994 when several PPAs with IPPs were signed. Far from the expectation that privatization would increase efficiency and reduce production costs, the experience with private investors has been that they have only added to the financial burden of PLN. This is because the involvement of the private sector was not accompanied by a requirement for competition and almost every contract was probably won not through a fair and competitive bidding process, but through inherited privilege and cronies of Suharto, Indonesia's then president. The requirement in the agreement obliged PLN to buy in US dollars 80 per cent of the capacity with a 'take or pay' system, which meant that PLN had to buy whether it needed the capacity or not, for the period of 30 years. This requirement was in the best interests of the private investors but not of PLN, which was pressured to sign the PPAs. Estimates of electricity demand were all but cast aside in the signing of 26 contracts, and over-supply resulted. If there had been no economic crisis, demand would have been approximately 10,500 megawatts and PLN would be able to meet demand with its own power generation facilities, which had a capacity of 14,000 megawatts. According to the World Bank, PLN traditionally built very large reserve margins, compared to peak load, into its generating capacity. The planning of private power generation also ignored the lack of transmission and distribution lines to carry the electricity to the consumers. Therefore it is of the utmost urgency that PLN re-negotiate the PPAs. With the fall of Suharto, PLN considered this move in the hope that the price of private electricity could fall to no more than four US dollar cents with the capacity approximately 30 per cent. However, US government officials and a number of US-based companies involved in the contracts have opposed this effort. Politically speaking, should Indonesia carry on with the renegotiation effort, it will ruin Indonesia's reputation for international investors.

Time to Restructure?

Given the current condition of the power sector, the government believes that this is the time to restructure, with the objectives of competition, transparency and more efficient private sector participation. The power sector would then be able to provide a better service to customers and support Indonesia's economic growth. This would be done by separating the functions of the electricity sector, reducing the government's involvement and opening the system to more players. This would mean unbundling PLN and establishing a system that operates outside the Java–Bali grid, with a separate company, the Regional Electricity Company (REC), being formed to manage this distribution. Electrification of this area is valued by government to boost local development and cannot operate solely as a commercial service, so the REC will be government-owned. Private participation outside Java–Bali would be limited to the contracting out of new projects for generation, transmission and distribution to private investors chosen by competitive and transparent bidding.

Unbundling PLN's functions into smaller units was planned to prepare these for later sale to private investors. Two new companies, the Java–Bali Electricity Company (JBEC) and the Java–Bali Transmission Company (JBTC) would be created in the middle of 1999. JBEC would be a holding company with several generation and distribution subsidiaries and would oversee the conversion of its subsidiaries into independent companies. Generators, including the existing IPPs, would compete with each other to sell electricity to the newly formed JBTC, which would then sell the electricity to distribution or retail companies and industrial consumers. This system, the 'single buyer', was to be established in 1999. Later on, the government would move to a 'multiple buyer/multiple seller' system in which generation companies could sell directly to distribution/retail companies through a 'power purchase pool'. The market would determine the price. Thus, a fully competitive market was hoped to be realized for the generation and retailing segments of the power business, while the transmission segment would remain regulated since it was a natural monopoly. Although the electricity charges would be determined through market mechanisms, the government would implement price caps and calculate allowed costs assuming efficient operations, for the basis of which tariffs and subsidies are set. The government would then reduce the allowable costs over time to promote efficiency and ensure that subsidies for poor consumers and less developed regions be made explicit and transparent. To assist this process, the government proposed to set up an autonomous regulatory agency to regulate the entire energy sector, public and private, and report to the minister of mines and energy.

The World Bank and the ADB prescriptions for privatization of the energy supply began to take effect in 1990 when the president approved, in principle, the allocation of portions of PLN for private sector implementation. A Private Sector Decree in 1992 established that the PLN's service area would be opened to private sector electricity generation. The multilateral lending institutions also stated their inability to meet PLN's projected financing requirements, and this prompted government recognition of an impending acute shortage of funds for expansion. In 1992, a ten-year investment plan was approved that would require an annual investment of $3 billion, roughly three-quarters of which would have to come from foreign sources. Because of the severe electricity shortage, the government pursued privatization without thorough assessment, which caused enormous problems in the areas of planning, project selection, and the bid solicitation and evaluation process. In both the solicited and unsolicited electricity plans, a project-by-project approach to private power implementation proved disadvantageous. Paiton One is a power project whose site, technology, fuel and scale were predetermined by PLN. Hidden subsidies in the forms of equity for system expansion, a low interest rate on lending, and absorption of foreign exchange risk were revealed when the industry was privatized. The imbalances in electricity pricing have shown that the pricing by the private sector is higher than the price that PLN has to charge the consumer. Because the president approves the pricing and the true economic costs of electricity are not known, the PLN has to subsidize the prices for the consumer and private sector. The monopoly system and the lack of transparency of the pricing regulation make the PLN responsible for the costing of electricity. The private sector has to be able to make operational an efficient generating system to reduce generating costs and make prices competitive in the market. It has to allow regulation to make electricity pricing more reliable and show the economic cost. Privatization should be more than just shifting the financial responsibility from public to private sector; it requires careful restructuring and deregulation of PLN and, most importantly, of the industry to allow a conducive environment for private investment.

In Indonesia, privatization offers a chance for efficiency and competition, which could eventually lead to lower costs and better services. However, in practice, it has often been unable to meet these objectives. It was hoped that privatisation would result in more transparency, so that at least the consumer would know how the tariff was fixed. The tariff would then closely reflect the true economic cost. Past experience has shown that many consumers did not know the tariff mechanism, however. In this case, although the cost due to inefficiency could be reduced, the tariff would be highly unlikely to be reduced since the present price is well below true cost. On the contrary, it would increase as the subsidy was gradually removed. Even though the price is expected to be determined by market, the regulatory agency would be in a strong position to decide the tariff. In the price cap regulation, the regulator would set the allowable price. Therefore it should be ensured that the regulating agency is really independent and that its members do not benefit from the increased tariff. Tariff setting should not be reduced to a mere technical calculation. An efficient price is not necessarily equitable. Value judgements would be unavoidably involved in decision-making in order to balance efficiency and equity.

As the theory goes, competition provides consumers with the freedom to choose products on the basis of price and quality and thus 'good' products would win the market and electricity companies would compete to bring ever-better services to consumers. But the freedom to choose is actually quite limited for small consumers of electricity, and it is the industrial consumers with high demand who buy directly from the power pool who can change electricity suppliers if they are not satisfied. So regulation of quality standards should be established to protect the consumers. The regulatory agency should also have a monitoring procedure to make sure that the regulation is enforced. Electricity companies that disobey the regulations or give poor service should be penalized.

Besai Hydropower Plant

Besai hydropower plant is under construction with PT PLN and World Bank finance to produce 90 MW of electricity. The total cost is estimated at US$206.67 million, around US$69.45 million of which is from local financing, with the remaining US$137.22 million from the IBRD. The project, which began in 1993, is aimed at regional development in the province of Lampung, Sumatra and will be fully operational in 2000. It will be a run-of-river power plant, with a daily regulating storage capacity. The powerhouse will be a surface type with two units of turbines and generators of 45 MW each. The total surface area to be acquired for the purpose of constructing the plant is about 202 ha, of which about 140 will be inundated, while the rest (62 ha) will be required for the dam, powerhouse, office complex and housing colony, contractors' use and access roads. Environmental impacts include soil erosion, air and water pollution from construction and waste disposal, sedimentation of the reservoir, scouring of the river-bed, risk of damages due to earthquakes and floods, loss of forest lands and wildlife, and an increase in water-related diseases. Apart from the impacts of the nearby construction, a total of 129 families will be affected, 88 of which occupy the area to be inundated. PLN has said that 41 families are categorized as forest encroachers with illegally planted coffee bushes on land belonging to the Forestry Agency. Conversely, a local NGO report claims that the families legally occupy the land and that approximately three hundred families in three rural areas are directly affected by the project. Rice-fields and plantations have been acquired and the local means of livelihood lost.

Allocated budgets are included for land acquisition and compensation, rehabilitation and monitoring of rehabilitation and land compensation. The Resettlement Plan has two main components relevant to the community; both are based on the assumption that the project management is committed to maintaining the community's quality of life, both physical and social. The mechanisms of resettlement and rehabilitation are implemented by the Land Compensation and Assistance for Rehabilitation of Project-Affected People (PAP). The land compensation committee carried out the project without PAP involvement. The rehabilitation project provides some vocational training, employment by the project, and counselling, but does not guarantee that the PAP will survive using their new skills in the rural area. The communities, including the PAP, know about the Besai hydropower plant plan from the village administrator. Before implementing the project, the World Bank did not approach the local community. Although PLN had provided information to the local community, on its own this does not constitute a genuine effort to involve them.

Environmental Degradation and Room for Improvement

The choice of technology and fuel for power generation facilities has implications for local, regional and global atmospheric pollution. Coal-fired power generation can cause acid rain. Fossil fuel-based technology emits greenhouse gases that contribute to global climate change. Indonesia has extensive regulations on environmental protection, with standards on air and water pollution. Environmental impact assessment is required for projects predicted to have adverse environmental impacts. However, enforcement of environmental regulation has been low, and environmental degradation is still occurring. Many environmental cases brought forward by people who have suffered environmental costs remain unsettled or are unfavourably resolved. Privatization of the power sector therefore would not be likely to bring any improvement in the environmental performance of the electricity business. Private investors' orientation is profit maximization, reducing the costs whenever possible, and it is unlikely that they would internalize environmental costs if not properly monitored.

From an environmental perspective, good practice in the power sector means utilizing clean power-generating technology and using energy efficiently by reducing losses and improving end-use efficiency. Privatization would make electricity bills reflect the real economic cost of the service. However, since environmental costs are not reflected in the market price, privatization might not favour 'clean' technologies unless there is some intervention from the government. If decisions are based solely on market mechanisms, the choice of technology and fuel depends only on the price of the technology and fuel in the market. The government could encourage the use of more environmentally friendly technology by creating some economic incentives and disincentives or some regulations in favour of the cleaner technologies.

Privatization might not support the development of renewable energy, which has relatively very small environmental impacts. The specific characteristics of most renewable technologies made them unattractive to private investors. Investing in renewables is considered to be more risky than investing in fossil fuel-based technology. The capacity and reliability of renewable energy depend heavily on nature and are not easily controllable. Renewables are site-specific, and consumers should be nearby or need adequate transmission and distribution lines. Renewables have a high initial cost, so need a relatively longer time to break-even. Private investors support efficiency only if the efficiency gains accrue to them. Supply-side efficiency would certainly be encouraged as it could reduce the production costs. On the contrary, end-use efficiency would be unlikely to be advocated by a private electricity company. Given the current situation of over-supply of electricity in Java–Bali, the assumption would be that there is no need for consumers to save energy. However, even at this stage the campaign to save energy is necessary. In fact, there is already a nationwide energy-saving campaign, but the response from consumers, especially in the urban areas, has been unsatisfactory.

The reality of Indonesia's economic crisis, human rights abuses, corruption, collusion, and nepotism means that even to discuss privatization, these fundamental problems must first be solved. If this is to happen, it will surely mean resolving many problems related to the Indonesian government.

Privatization may or may not be the best solution to the current problem in the power sector. In the case of Indonesia, it could be either beneficial or detrimental. The ideal scenario is that electricity companies be independently run, so that they can play their 'producer' role to a maximum extent. This will work, however, only if the government is able to leave behind its props and attributes as the owner of PLN. This is the only way to ensure that equity will be brought about, and is an essential condition for successful privatisation in Indonesia. Further, besides producers and policy-makers, an even more ideal scenario would be to create a neutral and independent regulatory agency. The regulator's function would include designing regulations, monitoring, and evaluating producers. Its neutral role has to be emphasized in order to accommodate the interests of all parties. The regulator could also play an important role in safeguarding the environment. So far, policy-making and regulation-making have both been kept well in the hands of the government. Another advantage would be to increase public participation in decision-making. This would create a reciprocal relationship between the government, regulator, producers and consumers.

Such an 'ideal' privatization could come about only within the context of democratic surroundings and a clean government. The overlapping governmental role, as the owner of PLN, and also as policy- and regulation-maker, should therefore be separated. Within the current situation in Indonesia, it would be beneficial to minimize government intervention. This would remove conflicts of interest. Democratic government and the independence of PLN would reduce instances of corruption, collusion and nepotism in governmental institutions closely connected with PLN, and would create fairer competition between producers.

Amid the current political uncertainties in Indonesia, goodwill appears to be present within the State Ministry of Mines and Energy, which has invited numerous NGOs and university-based research institutions to design a working group for a better management of Indonesia's energy sector. It is still rather early to judge to what extent this initiative will be successful, but it does have the potential to lead to more socially oriented energy management in Indonesia.

Note

1. This section was prepared by Agus Sari and Chrisandini, Pelangi, Janlan Danau Tonano, A-4 Jakarta 10210, Indonesia.

9

The Philippines[1]

Energy in the Philippines

To understand the Philippine energy sector, one must first understand its key player – the state. From the introduction of the Philippine Bill of 1902 up to the 1987 Constitution, 'All lands of the public domain, waters, minerals, coal, petroleum, and other mineral oils, all forces of potential energy, fisheries, forest or timber, wildlife, flora and fauna, and other natural resources are owned by the State.' Control of these resources occurs through agencies that implement government programmes in the energy sector and power generation sub-sector. The National Power Corporation (NPC) was established in 1936 to develop 'hydraulic power and the production of power from other sources'. With changes in the character of the energy sector and technologies available, the charter was revised 'to undertake the development of hydroelectric generation of power and the production of electricity from nuclear, geothermal and other sources, as well as the transmission of electric power on a nationwide basis'. The NPC was also tasked with the watershed management of sites near hydroelectric plants and given authority to 'adopt measures to prevent environmental pollution and enhance the conservation, development and maximum utilization of natural resources'. Of these revisions, the most significant was the nationalization of a state-owned power grid. NPC monopoly of the power generating sub-sector was institutionalized under the martial rule of then-President Ferdinand Marcos, when the NPC was tasked to 'own and operate as a single integrated system all generating facilities supplying electric power to the entire area embraced by any grid set up by the NPC'. In 1977 the Department of Energy (DOE) was established to 'further rationalise the country's total energy resource development program in order to accelerate its self-reliance and conservation ... relative to energy resources on an integrated and comprehensive basis'. The decree also foresaw a change in the energy sector; by recognizing 'the significant and continuing participation of the private sector in the various areas of energy resource development'. The DOE was abolished in the midst of a popular uprising that overthrew the dictator and regarded DOE as a vestige of the corruption that marked 14 years of dictatorship. In the early 1990s, the Philippines began to experience a massive power shortage that was largely attributed to the fact that no new power plants were built during the administration of Corazon Aquino, who succeeded Marcos. Newly installed President Fidel Ramos re-established the DOE 'to ensure a continuous, adequate, and economic supply of energy with the end ... of achieving self-reliance ... [and] to keep in pace with the country's growth and economic development and...active participation of the private sector in the various areas of energy resource development; and (b) to rationalise, integrate and co-ordinate the various program of the Government toward self-sufficiency and enhanced productivity in power and energy without sacrificing ecological concerns.' The private sector's role was further recognized when the DOE was mandated to 'endeavour to provide for an environment conducive to free and active private sector participation and investment in all energy activities', and to enact 'a policy direction toward the privatisation of government agencies related to energy, deregulation of the power and energy industry, and reduction of dependency on oil-fired plants.' The Philippine Energy Plan of 1996–2025 was prepared by the DOE in 1995 to articulate a relatively long-term policy perspective that would provide adequate, reliable and affordable energy to industries and to the ordinary citizen, while promoting sustainable development and maintaining the country's overall economic competitiveness. Encouragingly, the Philippine Energy Plan recognizes alternative sources of energy and the need to integrate environmental concerns into the planning and implementation process, but these policy pronouncements fall short in practice. For example, large-scale utilization of new and renewable energy sources is to be pursued, according to the plan, whereas in practice the new and renewable sources refer only to biomass such as rice and coconut residues, bagasse, wood wastes, animal wastes and municipal wastes. While biomass utilization is projected to increase over the planning period, its contribution as a percentage of the energy mix is expected to decrease to less than half its present level. Also, most of the biomass energy would be derived from wood waste and hence contribute to the release of CO_2. In terms of energy efficiency and demand-side management, the expected contribution averages only around 4 per cent of the total energy mix per year despite a stated policy to promote the judicious conservation and efficient utilization of energy. The roles of both the National Power Corporation and the Department of Energy are expected to change substantially in the coming years, partly because of legislative efforts within the country. The roles of the MDBs in effecting and influencing the direction of the changes, however, cannot be downplayed.

The Importance of the Banks

Because of the nature of the energy sector of the country, particularly the central and often omnipotent role of the state, funding for energy activities naturally comes from the main source of public infrastructure investments: lending from multilateral banks. Over a span of almost thirty years, the Asian Development Bank has provided US$1,766.2 million in loans and US$5.483 million in technical assistance grants. Almost a quarter of the World Bank's lending to the Philippines from 1988 to the present went to the energy sector. It was only in 1988 that the Bank resumed 'active' lending to the Philippine energy sector. The WB and the ADB have also influenced the policies governing the energy sector. Under the 'Philippines' section of its 1989 *Annual Report*, the ADB made very clear their intent to 'support private initiatives in the power sector in line with its overall Bank wide strategy to encourage build–operate–transfer (BOT) type projects.' Since then, the ADB has provided US$843.35 million in loans and US$3.275 million in technical assistance grants. These would account for around half of their total energy activities in a span of only six years. In their words, 'The widespread electric power shortage, which had brought growth almost to a halt in 1992–1993, has been tackled expeditiously by some innovative policy initiatives that enabled significant private sector participation in power generation.'

The World Bank has been more direct in its influence via loan agreements. The US$250 million 'Transmission Grid Reinforcement Project' was approved in 1996. The contention of the WB and the ADB is that the policies that were enacted by government to overcome the power supply crisis are the same policies that would answer projected energy demands. It can then be expected that the same issues and concerns raised against these 'emergency measures' would continue to be raised by the citizenry. During the power shortage, NGOs and the general public raised several issues that continue to be relevant. Of major concern have been the social and environmental impacts of power plants. In the past, power generation projects have been opposed by affected communities, partly because the government has been emphasizing large-scale projects with known socio-environmental impacts. And there is no change in sight with the government penchant for large-scale projects.

The Quezon Power Plant

The story of the Quezon power plant starts even before the project's inception in the early 1990s, when Luzon, one of three major island groupings in the Philippines, experienced power outages that resulted in blackouts in the capital, Metro Manila. As a result, a number of policy changes were undertaken, including the enactment of the law creating the Department of Energy (DOE), which had been abolished in 1986 by the post-dictatorship administration. Acute power shortages were the impetus for the government to accelerate privatization of the energy sector, particularly of the power generation sector, through the build-operate-transfer (BOT) scheme and its numerous variations. One of the first projects under the BOT scheme was a 2 x 350-MW baseload coal-fired thermal power plant in Pagbilao, an island off the coast of the province of Quezon, the southern part of Luzon. A Hong Kong-based company, Hopewell Holdings Ltd, undertook the project. The Asian Development Bank and the World Bank's private lending arm, the International Finance Corporation, provided a portion of the project's cost. The privatization theme was to re-emerge in 1994, this time under a variant of BOT, the build-operate-own (BOO) scheme. This time, it was the lives of the people of an agricultural community, the municipality of Mauban, that were to be affected.

A visitor to Mauban once remarked that 'Many discover the South, some even make it to Mayon [a volcano in the Bicol region famous for its near perfect conical structure], but only a chosen few discover Mauban.' The 416-square kilometre coastal municipality had once been surrounded by forests, but illegal logging and deforestation had defaced the surrounding hills. It was this destruction that caused the community and church organizations to monitor and protect their environment. The primary occupation of the Maubanens was coconut farming, and not a few were beneficiaries of past agrarian reform programmes. This, however, was to change in 1994 when the Quezon Power Plant, a 433-MW coal-fired thermal plant, was proposed for construction. Quezon Power Ltd, a partnership between a US company, Ogden Power Corporation, and a Philippine corporation, Pacific Manufacturing Resources, undertook the project. The partnership was obviously formed to circumvent the constitutional prohibition on the exploitation of natural resources, including energy sources, and public utility franchises, by corporations less than 60 per cent owned by Filipinos. In addition, Bechtel Enterprises, another US-based firm, won the construction contract. Quezon Power entered into a power purchase agreement with the country's largest electric distributor, Manila Electric Company, owned by the country's leading oligarchies, the Lopez family, with interests in water supply, mass media and communications. From its inception, the agreement was suspect – the company had asked the governor and representatives of the province to recommend the project to the Department of Energy, citing the economic windfall that would subsequently befall their province. In a country where political patronage is the key to obtaining government approval, the project was easily endorsed, despite opposition from the community in question.

As required by law, an Environmental Impact Study (EIS) was prepared before the project could be given an Environmental Compliance Certificate (ECC). BHP Engineering, a foreign firm with an office in Manila and only a tentative grasp of the Philippines environment, conducted the EIS. To describe the existing situation in the Probable Impact Zone, the EIS stated that a 'typical house is made of *temporary* materials such as cogon, nipa and/or bamboo' (emphasis added). Even first-time visitors to the country would notice that houses made of these 'temporary materials' dominate the rural areas. Such houses, called the *bahay kubo*, have long been considered symbols of Philippine culture, and are considered to be architecturally and structurally adapted to the country's climate. Comparisons between a *bahay kubo* and concrete row houses have been made a number of times in the past, however, to show how communities affected by power projects end up in 'better' houses in resettlement areas, for the sole reason that these are made of concrete. It seems to matter little that the former are cooler and more airy, and in most cases larger: the use of concrete is a powerful symbol of 'development' and 'progress'. Two factors were added to sum up the social impacts in the EIS: income levels and poverty incidence. It is a paradox of the EIA process that social acceptability is included as one condition in the permit to undertake environmentally critical projects, because the mere possession of such a certification then gives the proponents the right to start the project, leaving affected communities with no choice but to give 'social acceptability'. Once the proponent is issued an ECC, there is virtually no room for the community to stop the project, as consultations at this stage become merely ceremonial. The ECC, too, in practice affords no protection of affected communities. The government agency mandated to ensure compliance has never shown that it has either the teeth or the spirit to fulfil its mandate. As a response to growing community protests in Mauban regarding violations of ECC conditions, in 1997 an official task-force was sent to investigate and a Cease and Desist Order was issued ordering the suspension of operations and the payment of approximately US$6,200. The community, while happy with the temporary victory, had been hopeful for nothing less than the total revocation of the ECC, for clear breaches including entry into private property and destroying trees without owner consent, health and safety hazards due to hauling operations, improper disposal of solid and liquid wastes from the plant site, failure of the proponent to provide effective mitigating measures against dust generation and deterioration of road surfaces, and the proponents' failure to secure the appropriate permits for water extraction. The community argued that these five violations were mild compared to the other violations that had occurred, but the DENR nevertheless imposed a fine of US$6,200 and lifted the Cease and Desist Order (CDO) less than a week later, upon the proponent's promise to do good.

The task-force reported findings ranging from the accurate to the ludicrous. 'As an immediate solution to the health complaints, perhaps the proponent should consider providing residents with dust masks, especially for children, to prevent dust inhalation.' The very important complaint of damage to water systems was referred back to 'the MMT [Multipartite Monitoring Team] and the proponent to investigate the matter and act accordingly'. The MMT, particularly its composition, was itself a concern that was raised. In effect, the task-force passed on the responsibility of further investigation to the culprits. Some violations, however, did not escape notice. The Environmental Guarantee Fund (EGF) Committee, one of the mechanisms highlighted by the government as a means of community participation, was found to contain an employee of the proponent who had been listed as an NGO representative. Because of the unanswered questions and the premature lifting of the Cease and Desist Order, the Mauban community again requested the Department of Environment and Natural Resources (DENR) to conduct an investigative mission, which it finally did in July 1998.

During the February 1999 confirmation hearings it became clear that the issue of ECC violations in Mauban had been maliciously withheld. Unfortunately, the violations have continued. One of the more contentious issues raised by the community is the cutting down of trees without required permits. Philippine laws protect certain trees from indiscriminate felling, and permits are required from the relevant authorities before they can be felled. Further, hardwood species, particularly narra and batino, were reported to have been cut down on three parcels of land. The investigation teams, however, to the consternation of the community, overlooked these clear violations of the law. In order to facilitate acquisition of the almost hundred hectares needed for the plant, Quezon Power (Philippines), Ltd (QPL) organized Mauban Properties, which, even before information about the power plant was released, began negotiating with farmer landowners. Their ranks were systematically decimated until only a single landowner, Concordia 'Tia Cording' Malubay, refused to sell out. In November 1997, QPL filed several complaints in order to enforce right-of-way (ROW) agreements allegedly entered into by several landowners of lots surrounding the actual construction site, which was to be used for transmission lines to connect the plant to the Luzon grid. The farmers asked for the dismissal of the cases on several grounds, including the failure of QPL to state a cause of action. Realizing its mistake, QPL also asked for the dismissal of the cases. Just a week after the cases' dismissal, however, Meralco filed a case for expropriation against the farmers before the Regional Trial Court (RTC). The power to expropriate land with due compensation to the owners has been delegated to Meralco, a private corporation. In the Mauban case, Meralco wants to exercise this power to benefit QPL, a transnational corporation, over the farmers. Despite the farmers' motions to clarify Meralco's petitions, the court, without holding any hearing, issued writs of possession to Meralco covering the entire lots of the farmers. The Mauban farmers were surprised that the court, which has not installed computers, was able to issue *computerized* writs of possession. It was on the basis of these bogus writs that the sheriff and armed men of the Philippine National Police (PNP) 417 Mobile Group, Polo Detachment, along with QPL personnel, entered the lots and built a series of transmission towers and access roads. After repeated refusals by the court to quash the defective writs, the farmers brought their petition to the Court of Appeals on 6 January 1999. The RTC of Mauban fast-tracked the proceedings and summarily condemned the property of agrarian reform beneficiary Miguel Encina on 21 January 1999. The land involved is prime agricultural land used for the construction of transmission lines.

There have been a number of controversial energy projects where the World Bank or the Asian Development Bank have been involved. Chico River dam, a project in Northern Luzon funded by the World Bank in the 1970s, faced stiff opposition from the indigenous peoples who were to be displaced. Different indigenous groups in the Cordilleras united to fight the project and the World Bank eventually pulled out, but not before instances of violence, including the death of a tribal leader, Macli-ing Dulag, due to the intensive militarization of the area by the government, which was determined to pursue the project. The Chico River dam struggle has been one of very few successful campaigns against developmental aggression in the Philippines. Mt. Apo Geothermal Plant has the highest peak of the Philippines and is home to numerous *lumads*, or indigenous peoples. Mt. Apo had been declared a national park, which would have automatically disallowed any power generation project in the area. But *lumad* groups were forced to enact a *d'yandi*, or pact, to fight the project to 'the last drop of their blood'. The WB and ADB eventually pulled out of the project, but funds were sourced through other institutions. The Agus 1 Hydroelectric Plant, a dam project in Lake Lanao affecting the Maranaws ('people of the lake'), has led to alternate drying and flooding of the banks where the Maranaws live. The ADB provided funding for the transmission lines of this project. Casecnan Trans River Basin, a multi-purpose dam project in Central Luzon affecting the Agtas, another indigenous group, has been stalled by the opposition of various groups, including the New People's Army. But the government has pursued the dam construction primarily because funds are available, partly from the ADB. Pagbilao Coal-Fired Thermal Power Plant in the southern Tagalog region of the country displaced communities and their agricultural lands. Those living near the plant complain of environmental degradation caused by the plant's operations. This plant is funded by the ADB. Masinloc Coal-Fired Thermal Power Plant is a 600-MW plant in central Luzon. Farmers and fisherfolk living in and near the plant opposed it because it would mean economic dislocation. The opposition, which used to include even local government officials, was weakened after a steady stream of funds from the government divided the people. The project was stalled for almost eight years, but the ADB never reconsidered its commitment to fund the project, despite campaigns on local, national and international levels.

Changes Afoot?

In response to numerous experiences with affected communities, the NPC has adopted several changes. It cannot be said at present that the NPC totally disregards the sentiments of peoples affected by the power plants. In fact, it is because of this recognition that the NPC has instituted changes in its methods of establishing a plant. Community organizers are now deployed, promising economic benefits in exchange for acceptance of the plants. Pictures of model resettlement areas are shown during public meetings. In these meetings, the bulk of the presentations are about the resettlement package, and not about the actual project and its impacts. As in the past, there is no option for non-continuance of the project.

Second, there is still a bias towards fossil fuel-based power generation. As it is, coal-fired and oil-fired plants dominate the power generation subsector. And the prevalence of fossil fuel-based technologies is to increase. According to the Philippine Energy Plan, a total of 92,138 MW will be added by 2025, of which 39,050 MW will come from coal-fired plants, and 30,385 MW will come from oil-fired plants. This represents 75 per cent of the expected new installations, and 700 per cent of the present power generation level for all sources.

Further, power development is intricately linked to the overall development and industrialization plan that continues to be questioned and opposed by the citizenry. The energy requirement projections assume the growth projections of the Medium-Term Philippine Development Plan 1993–1998, which attempts to fast-track industrialization by increasing foreign investments at the expense of sovereignty, and building an export-oriented market at the expense of national subsistence. Questioning this fast-track newly industrialized country (NIC) vision should, similarly, remain central to any energy advocacy; any other approach and method would be merely palliative.

The investments needed for the energy sector in the next 15 years amount to around US$71 billion, with the private sector expected to contribute 83.1 per cent of the total. For the next 30 years, on the other hand, a total of around US$297 billion is required, 90.9 per cent of which would come from the private sector. There is still a role, therefore, for the investment of public funds, and the WB and ADB are expected to continue to be active in the following areas:

- direct loans for transmission projects (public sector)
- private sector loans under BOT (or its variant) schemes
- technical support for structural changes in NPC
- technical support for energy efficiency and DSM initiatives
- development of hydroelectric power, particularly in Mindanao (though not in the immediate future)

With reference to the second point (on private sector loans), the WB and ADB should include private sector activities in the scope of their policies and regulations of information disclosure, participation, social acceptability, environmental impact assessment, and accountability. Requirements for information disclosure and accountability must go further than present MDB standards, given that private sector entities are often unfamiliar to affected peoples and communities. Add to this the fact that loans to the private sector are processed twelve times faster than public sector loans, and take two or three months as opposed to two to three years. As it is, information is not disclosed even after the loan has been granted. However, it is preferable still that the MDBs desist from destructive energy sector activities in the Philippines and this is currently occurring, albeit for the undesirable reason that they have scaled down their operations to encourage a greater private sector role in the power generation sector. The MDBs could provide technical and financial assistance for the development of alternative energy sources and to institutionalize energy efficiency, but instead these remain side issues to their main agenda: privatizing and commercializing the energy sector.

The WB and ADB could even pick up the tabs for the US$2.4 billion projected requirements for demand-side management (DSM – see Box 2.3), energy efficiency and other environmental activities over the next 30 years (around US$80 million a year – considerably less than the US$100 to 250 million usually provided for a single power generation or transmission project). But in the greater picture, large-scale traditional power generation and its impacts would still overshadow contributions to the energy mix and environmental mitigation from these activities.

Note

1. This chapter was prepared by André Ballesteros, the Legal Rights and Natural Resources Center, 3/F Puno Bldg. #47 Kalayaan Ave., Diliman, Quezon City, Philippines.

10

Bulgaria[1]

Energy Shortages and High Import Bills

While under Soviet control, Bulgaria followed a very energy-intensive development path, with heavy dependence on fuel imports and Soviet-designed power plants – most of which have low safety, environmental and efficiency standards. The collapse of communism in Bulgaria in 1989 had serious consequences for the energy sector. The combination of the unstable political and economic situation in the country, lack of experience with a market economy and a disparity between existing industries and the competitive international economy created a financial crisis for consumers and industry alike, which led to a significant decrease in the demand for electricity. This decreased demand has nevertheless failed to prevent widespread energy shortages and high energy import bills.

In response to the crises facing the country's energy sector, the Bulgarian government has partially liberalized petroleum fuel prices and sharply raised the prices of other energy goods. The government has also initiated a least-cost programme and several studies on the rehabilitation and safety improvement of the power system. A number of key issues remain unresolved, however, such as: inadequate organization and regulatory set-up of the energy sector; relatively unstable electricity supply; high electricity costs; highly inefficient supply links; inefficient end-use; an urgent need for rehabilitation of a number of power plants; and general nuclear safety and waste concerns, particularly in relation to the Kozloduy nuclear power plant.[2]

In general, there are no legal incentives for energy efficiency in Bulgaria. Legislation in the nuclear energy field is also woefully insufficient: safety procedures in the case of accidents are lacking, and there is a lack of clarity concerning which organization should take responsibility for the radioactive waste from Kozloduy.

Bulgaria has received substantial financial support from the MDBs for its energy sector. There are five projects under implementation financed by MDBs in Bulgaria totalling US$247.2 million, and five more projects in the pipeline, totalling US$675.5 million. The loans are directed mainly towards rehabilitation, upgrading and improving the efficiency and safety of the district system, and for the Maritza East II coal-powered plant, the Chaira pump storage plant, the Belmeken and Chaira dams, and the Kozloduy nuclear power plant. A more detailed description of the World Bank's largest energy investment in Bulgaria, the 'Energy I' project, follows.

The World Bank's 'Energy I' Project

This World Bank-financed project was initiated by the Bulgarian government in 1991 and sponsored by the National Electricity Company of Bulgaria (NEK). The expected completion date of 31 December 1996 was extended – due to 'problems with project implementation' – to 30 December 1998. Total project costs are estimated at US$126 million, of which the International Bank for Reconstruction and Development's (IBRD – a component of the World Bank) contribution was US$93 million. The remaining US$33 million was invested by the NEK.

The project's objectives are: to improve the operating efficiency and reliability of the system; to reduce the need for electricity imports to meet peak demand; to realign electricity tariffs to rationalize consumption, reduce imports and pollution (including CO_2 emissions) and mobilize resources for the NEK; to improve voltage and frequency control; to enhance the organizational efficiency of the NEK and to 'depoliticize' the process of setting energy tariffs; and to improve safety at the Belmeken and Chaira dams.

The project is separated into several components, which are oriented towards: improvement of the supervisory control and transmission systems of the electricity grid through installation of improved communication equipment, advanced computer software and training, and improvement of grid reliability; completion of the already 90 per cent completed units 3 and 4 of the Chaira pumped storage plant (PSP), along with strengthening of the associated dams in order to provide an additional 432 MW of peak capacity; provision of technical assistance for reorienting the NEK's operations along commercial lines; improvement of financial management, establishment of a modern accounting system and computerized management information, billing, collection and cash management systems; and building the skills necessary for hydroelectric dam monitoring.

As an aside, it should be noted that the basic idea for completing the Chaira power plant had been to have a stable peak-time working power plant to avoid the safety problems associated with running the Kozloduy nuclear power plant at full capacity during peak times. It may be possible, however, that after the completion of units 3 and 4 of the Chaira pumped storage plant, energy demand will be stimulated, leading to a need to complete the nuclear power plant at Belene. This means that the project may lead to a 'through the back door' increase in nuclear power in Bulgaria.

The loan's main conditions are: maintenance of average electricity prices at 3.5 US cents as an interim measure; establishment of an independent price-setting mechanism and maintenance of the price at the level of long-run marginal costs; the NEK will gradually increase its provisions for depreciation to be in line with its revalued assets by the end of 1997; implementation of a set of measures for the NEK to improve its financial situation and financial management.[3]

The expected project benefits are: lower costs and increased stability of the electricity supply, resulting from an increased capacity for active load management, automatic generation control, and reduced losses; improved economic and managerial efficiency of the NEK; increased dam safety, as well as a slight reduction in pollution related to electricity generation.[4]

The project addresses some of Bulgaria's basic energy problems. During the project's development, the completion of the Chaira pumped storage plant, which was initially treated as a means of avoiding energy shortages, became part of a possible programme for the rehabilitation and increase of electricity capacity, to replace the first four units at Kozloduy. This is why the main components, listed above, focus on these problems. Other components focus on the rehabilitation of the electricity distribution system and overall grid management.

The above-mentioned delay in project implementation was caused by a lack of willingness on the part of the Bulgarian government to increase electricity prices to 3.5 US cents per kWh. This is part of a larger problem involving energy pricing. The Bank argued that the suggested price of 3.5 US cents would encourage the efficient use of energy and lead to increased capital investments. To date, energy prices have been largely determined within an obscure political process in Bulgaria. The so-called 'depoliticiza-tion' of electricity tariffs is one of the World Bank's main objectives in the country.

None of the Bank's principles for price reform was implemented after mid-1996, however, when the government adopted new higher prices. This is because it became clear then that such price 'reform' would only put the money into an economy of impoverished people with no real, functioning market and inadequate legislative incentives and restrictions. Rather than taking the opportunity to deal with the lack of open and transparent regulation in the country, however, the Bank substituted real market reforms and de-monopolization for a centrally planned administrative 'creation' of the market. Instead of supporting the development of independent power production in Bulgaria, therefore, the World Bank has in fact strengthened the NEK's monopolistic position and highly inefficient and opaque organizational structure.

Indeed, lack of transparency has characterized the whole project and has contributed to its problems. Only one section of the Chaira power plant completion project was evaluated by an Environmental Impact Analysis, and even this process involved very poor public participation.

Further problems resulted from the project's delay. The completion of the Chaira pumped storage plant represents part of the capacity that will replace old units 1–4 at Kozloduy. Any delay in the completion of units 3 and 4 of the Chaira plant means a postponement of the closure of the old units 1 and 2 at Kozloduy. The lack of a clear capacity-replacement strategy from the Bulgarian side has already led to an EBRD freeze on energy sector projects in Bulgaria. The World Bank's principles will also be put to the test on this issue.

In general, it is possible to draw the following conclusions:

- The significant delay in project implementation due to Bulgarian non-compliance with some aspects of the project exposes the problems of unaccountable state control of the energy sector, and underlines the need to de-monopolize and limit the sector's functions in Bulgaria. The project clearly illustrates the problems resulting from the NEK's monopoly and shows that this type of 'closed monopoly' is not working in Bulgaria, even with financial support.

- Some of the objectives of the project have been reached or will be met by the time that the project is completed. They are: (i) reduction in the need for electricity imports to meet peak demand; (ii) improvements in voltage and frequency control; and (iii) safety improvements at the Belmeken and Chaira dams. Other objectives, however, such as enhancement of the organizational efficiency of the NEK and the depoliticization of the setting of energy tariffs, cannot have significant and continuous effects, because of the NEK's position as a state monopoly. Indeed, the World Bank's objective to 'enhance the organizational efficiency of NEK' has proved to be at variance with the objective of 'encouraging independent power production'. This last goal cannot be achieved by strengthening the NEK, which is highly inefficient and non-transparent in its procedures. Moreover, because the NEK is currently included in the Bulgarian government's list of state organizations protected by security legislation, the public cannot obtain any official information from them at all, including figures on how MDB loans have been spent.

- The project has raised the question of whether Environmental Impact Assessment procedures should be fully completed before or after ratification of MDB investments by the Bulgarian parliament.

- It is doubtful that the project will be successful from the point of view of the general Bulgarian energy situation. There will be additional peak capacity, but the problems associated with the Kozloduy nuclear power plant will remain. Only an unambiguous agreement between the Bulgarian government and donors, supported by programmes and projects for investments in energy efficiency, energy conservation, renewables, and rehabilitation of existing hydro- and thermal power plants, monitored by open and transparent regulatory structures, will reduce the risk of a nuclear accident in Bulgaria.

- The project is not fully consistent with principles of public involvement. Except for some public discussion under the EIA procedures, there were no other areas of the project where people other than governmental officials and NEK staff were involved. Insofar as the project is part of a so-called 'set of investments' that the government should make in order to replace Kozloduy's old units, the participation of independent experts and other interested groups should have been treated as indispensable. As it happened, insufficient information was made available to the public during the entire project preparation and implementation period.

Price Reform is the Stumbling-Block

The World Bank's *Strategy for Bulgaria* gives the power and energy sector an important role: 'The Bank's involvement will mainly be directed at: … selective support for the development of power and energy and municipal infrastructure.'[5] The Bank's overall objective is to assist with financing the Bulgarian electricity sub-sector's medium-term core investment programme. This in turn is designed to enhance Bulgaria's non-nuclear electricity supplies at the lowest cost while reducing associated pollution, as well as facilitating gradual retirement of the four older units at Kozloduy.[6]

The major thrust of the Bank's strategy is energy price reform. This has become a stumbling-block in negotiations between the Bulgarian government and the Bank – and other MDBs – in relation to energy sector investments. The pricing mechanism promoted by the World Bank did not relate closely to Bulgaria's inflation rate, dramatically devalued the currency, and proved to be inadequate for the country's current unstable economic and political situation. The mechanism depended on political intervention and required formal governmental approval. Moreover, the commission set prices for energy state regulation without public knowledge or input. Even now, there is still officially no information available to the public on the pricing mechanism. This is despite the fact that a special commission responsible for the price mechanism has been established, headed by Bulgaria's trade minister, which proposes energy prices to a council of ministers, who then vote on the commission's proposals.

The difficult economic and political situation that Bulgaria has faced for the last five years has had a major influence on the process of initiating and implementing projects by international financial institutions (IFIs) in the country's energy sector. Some of the main objectives of the loans currently under implementation have not been fully met.

These unfulfilled objectives include the de-monopolization and restructuring of the energy sector; real energy pricing reforms; the retirement of the first four units at the Kozloduy nuclear power plant (NPP); and assisting the development of alternative sources of energy. Little has been done to promote demand-side efficiency, or the use of energy-efficient and renewable energy technologies. In addition, there has been in general very little public consultation regarding the present IFI projects. These unfulfilled objectives should be addressed in planning future IFI-funded energy-related projects in Bulgaria.

Notes

1. This analysis for Bulgaria was prepared by Petko Kovatchev and Ralitsa Pana-yotova, Centre for Environmental Information and Education, Jk. ILINDEN, Indje voivoda str., Bl.9 (61) entr.3, 2 floor, 1309 Sofia, Bulgaria.

2. In addition to the general problems associated with the old-type reactors at units 1–4 of Kozloduy, a major accident occurred in September 1992 at Unit 6 during reactor tests.

3. For example, after 31 December 1992, the NEK must establish funds from its own resources at a rate of not less than 30 per cent of its yearly average capital investments made or expected to be made during the present and the next fiscal years. Thus, starting in 1993 and for every subsequent fiscal year, the NEK's net income has had to equal 1.5 times all debts served during the same year.

4. *World Bank Lending Program*, updated March 1997.

5. EBRD information: *Strategy for Bulgaria 1996–97: A Summary*, 'Operation Objectives', section (iii).

6. *Bulgaria – Energy II, PID BGPA8321*, 4 March 1997.

11

Hungary[1]

A High Degree of Uncertainty

The current state of Hungary is a good illustration of the difficulties facing transitional economies. The long process of transforming the social-economic system involves a high degree of uncertainty, and changes are as likely to take a positive as a negative direction in a number of areas. Hungary has scarce mineral resources, so more than half of its primary energy demand has had to be supplied by imports, and energy policy has concentrated primarily on ensuring a safe energy supply. This was fairly easy under COMECON. Natural gas and crude oil came smoothly through the 'Friendship', 'Fraternity' and 'Adria' pipelines, and Hungary was also spared the impacts of oil crises. While local resources were explored and exploited, they played a relatively minor role in overall resource allocation. One exception to this is coal-mining, which until recently played a major role in employment policy. Therefore, the recent sudden abandonment of mining has proven extremely difficult both economically and socially.

As the recent political and economic changes came, Hungarian energy policy shifted towards the free market. Hungary was the first country in the region to start the long process of creating the legislative and institutional background for privatizing its energy industry. In the spring of 1994, the parliament passed bills regulating the gas and electricity industries, and the Hungarian Energy Office was established accordingly.

Hungary's electricity privatization legislation generally followed the British model, albeit in a completely different social and economic environment. At the same time, there were some crucial differences between the Hungarian and the British approaches, one of the most important being how access to the grid is managed. While in Britain, distribution utilities have free access to the supply grid and are allowed to sell electricity all over the country, in Hungary, regional electricity distributors control their own territory. This results in the maintenance of regional distribution monopolies. The second major difference is that under British privatization, the new owners of the energy utilities were at least initially British; in Hungary, they are foreign investors.

Generally speaking, the electricity industry in Hungary has been vertic-ally segmented. That is, production, transmission and distribution activities became separated, with the aim of making the electricity sector transparent. The sole exception to this is the Paks Nuclear Power Plant, which alone provides 43 per cent of Hungarian electricity production. This plant is still a part of the Hungarian Electricity Works (MVM Rt.) and forms one economic unit with the high-voltage grid. Currently, the installed capacity for electricity production in Hungary is 7,200 MW.

The gas industry was segmented into regional distribution companies and the rest (exploration, importing, storage, transmission and trade) remained with the Hungarian Oil and Gas Company (MOL Rt.), which is one of the biggest companies in Central Europe. MOL Rt. controls the Hungarian transmission capacities (pipes), although other companies are also allowed to import oil and gas into the country, since there is relatively little to export.

In late 1996, the Hungarian parliament passed a new nuclear law. Essentially, it makes possible the privatization of the already existing nuclear power plant in Hungary and allows private entities to build new nuclear facilities in the country. Moreover, the law establishes a nuclear fund for long-term waste problems and decommissioning. Unfortunately, this nuclear fund is insufficiently regulated. There is no regulation regard-ing the timing of cash flows, and there are no figures for the total funds needed at the end of the nuclear power plant's lifetime. The management of the fund is similar to that of other state funds, which are subject to the annual state budget. Considering these problems, it is doubtful whether the fund will fulfil its mandate.

The government is currently planning to start debating a district heating bill, although this has been an issue for years. Since the electricity bill does not clearly regulate the co-generation plants that currently operate through-out Hungary as district heating plants, the short- and long-term economic situation of these plants is extremely insecure. The lack of a clear regulatory scheme means that these plants sell electricity on a day-to-day and rather uncertain basis.

To date, the majority of the shares in gas and electricity distribution utilities have been sold to foreign companies (including Electricité de France, Bayernwerk, and RWE.) The generation capacity is only partly sold, as many of the power plants are very old (reaching the end of their useful life).

The only parts of the electricity industry in which the state still holds majority ownership are the high-voltage grid, three coal-fired power plants in the western part of the country, and the Paks Nuclear Power Plant Company. The gas industry is almost fully privatized. The majority of shares in the gas distribution companies are owned by German, French and Italian utilities.

Official Hungarian energy policy ostensibly favours energy efficiency, but at the same time, 4,000 MW of new capacity has been planned to 'replace' 3,000 MW of aged capacity by 2010. This does not reflect least-cost planning requirements. Officials claim that this is just an 'exception' now, because they have to hurry to secure capacity for the near future.

New legislation on the public participation process is under development in Hungary. This will have an effect on new power plant construction and other energy-related projects as well. This is currently an area of very weak regulation, as it does not provide for public involvement in environmental decision-making, but only informs the public about the possible effects of decisions regarding proposed projects and their environmental impacts.

Sustainable Energy Left on the Shelf

In late 1995, the Hungarian government approved the Energy Saving Action Programme (ESAP).[2] This document is based on a long study, prepared by the Ministry of Industry and Trade under the same title in 1993. The ESAP represented an official acknowledgement that potentially 25–30 per cent of energy can be saved, and energy sector energy efficiency can be increased, by efficiency improvements to power plants, reduction of grid capacity loss, and related energy conservation measures.

Heavy industry in Hungary has almost totally collapsed. At the same time, however, the share of residential and public consumption has been increasing. The overall result, therefore, has been an increase in energy intensity in the region. Moreover, it is the consumer and residential sectors that are most likely to grow in the near future; this is alarming because it is the industrial sector that is most likely to take energy efficiency measures, leaving huge potential energy savings in the agricultural, public and residential sectors.

According to the government-approved ESAP, 'If the necessary conditions are created, the current 2.5–3 per cent proportion of renewable energy resources can be increased to 10 per cent.' There is a strong potential for the use of thermal solar energy in public and residential buildings, as well as for the use of passive solar techniques. There is also a huge potential for the use of biomass and geothermal energy sources. Wind in Hungary is not strong, regular or frequent enough to use it for mass electricity generation, but small-scale wind power could be used locally in some places. There are also possibilities for small hydro projects in certain areas. Taking advantage of all these potentials could result in a higher share of total renewable energy consumption for Hungary, but unfortunately, no study has yet been carried out to prove this.

Despite legal authorization, the government has not appeared keen to develop the energy efficiency potentials described in the approved ESAP, which regards integrated resource planning (IRP) and demand-side management (DSM) as key elements (see Box 2.3). The plan will remain just another report on the shelf unless an appropriate financial environment is created for its implementation, and detailed guidelines are formulated for new owners of the reorganized electricity system.

The World Bank has approved several energy loans in Hungary since 1989. These include a US$100 million loan to the gas industry (MOL Rt.) for gas grid development, and a US$100 million energy and environment loan to the Hungarian Electricity Works.[3] Forty per cent of this loan is for the creation of a communication and information system within the Hungarian Electricity Works. Another 40 per cent is for the completion of an 80-MW gas-fired combined cycle power plant (Dunamenti G–II.). The remaining 20 per cent will be used for the development of an environmental plan, for pollution monitoring systems, for grid efficiency, and for 'educational purposes' for the management and other employees of the Hungarian Electricity Works (MVM Rt.); a US$60 million loan for gas turbines;[4] and a US$1.2 million feasibility study for three cities in the western part of Hungary on the potential of biomass. The Bank is developing the project, but the finance is coming from several governments, including those of Switzerland, Austria, Japan, Germany and Denmark. If this pilot scheme is successful then the Bank is likely to support similar schemes in other Hungarian cities. This will come, however, with a relatively large contribution from the GEF (Global Environment Facility). Investors in the project have made it clear that their interest is in the development of large-scale biomass projects, although local opinion favours small-scale schemes. In fact, local opinion has been largely ignored to date and, although an official process of public involvement and participation in the project is in place, the Hungarian public have been kept mainly on the outside.

The World Bank's private lending arm, the International Finance Corporation (IFC) has signed a contract with the UNIC Bank for an Energy Efficiency Guarantee Fund in Hungary. This is one of three future sources from which private companies can obtain guarantees for energy efficiency projects that would not be possible without this type of support.[5] The Energy Efficiency Guarantee Fund is financed by the GEF and provides only for leasing contracts. The IFC has established an advisory board for the fund, which includes some NGO representatives in addition to officials from related government ministries. This board does not have a strong influence on decisions but is used to solicit new ideas and as a channel to promote the fund throughout the country.

The European Bank for Reconstruction and Development has approved three energy-related projects in Hungary to date. These are: US$24.7 million for the MOL Zsana Gas Storage project, which aims to assist the Hungarian Gas and Oil Company to establish greater gas storage capacity; US$5 million for Prometheus Energy Service Company (ESCO) financing, in support of demand-side management (see Box 2.3) – a promising initiative that has come out of the EBRD's energy efficiency unit; and US$37.8 million for a general-purpose credit line for energy efficiency. Some of the conditions of this loan are unclear, particularly in relation to which organizations in Hungary will be responsible for different aspects of the project. This credit line is connected to the European Union (EU) grant programme for Central and Eastern European countries (PHARE), which has opened an energy efficiency revolving fund in Hungary. Typically, both the EBRD and PHARE have handled the project in a secretive manner and it is nearly impossible for the public to obtain relevant information.

The European Investment Bank (EIB) has also been involved in energy-related loans in Hungary. The EIB provides funding for energy efficiency projects only through so-called global loans: multi-sectoral loans that can be used for industry, tourism, energy efficiency and environment-related investments. There are no other types of energy efficiency-related pro-grammes supported by the Bank. Global loans are disbursed by commercial banks. This means that information on loans is generally secret and it is not possible to tell the number of projects or the total amount spent on energy efficiency through global loans. Energy efficiency is only one of the possible destinations for global loans, moreover, and it is safe to say that there is a low chance that any EIB global loans have been spent on energy efficiency projects. A total of US$44 million has been loaned to the Hungarian Electricity Works for grid rehabilitation and for feasibility studies on grid connection to the Western European system; US$44 million for rehabilitation at the Kelenfold Thermal Power Station; and US$44 million for a new gas turbine power plant on the site of a former power plant at Lörinci.

The World Bank's Key Role

The development banks, particularly the World Bank, played a key role in the legislative changes towards privatization of the Hungarian energy sector that began in 1993. This process was pushed through in Hungary at great speed mainly for budgetary reasons. As a consequence, when privatization began, a transparent and open regulatory framework was not in place. This created an uncertain market for investors and a critical social and political situation. Privatization of state-owned gas and electricity utilities began in 1995, after a long series of heated negotiations. Interestingly, German companies, such as Bayernwerk and Ruhrgas, and French companies, such as Electricité de France and Gas de France, seem to have divided their investments on a regional basis within the country, each taking over separate regions. It is premature to draw conclusions on this, but it may have important implications for the future development of both the Hungarian energy sector and the economy as a whole.

Generally speaking, although MDBs have supported a number of initiatives on energy efficiency in Hungary, the amount allocated by them for increased generating capacity or other types of supply-side projects is still much greater than that allocated for energy efficiency or renewables. It is also the case that even when the MDBs have been involved in 'environment-friendly' energy projects in Hungary, they have shown a general lack of openness and transparency.

Notes

1. This analysis was prepared by Ada Amon, Energy Club, P.O. Box 411, H-1519, Budapest, Hungary. A full length report is available in Hungarian. A more detailed analysis in English appears in *Time For a Change*, which is available via www.bothends/Mdb

2. Government decree: 2399/1995, XII. 12.

3. #3705 – HU/A, 1994.

4. #HUPA45251.

5. Another potential source is likely to be the IFC itself, and the other leasing company, INNOTRADE, has been under assessment.

12

Lithuania[1]

Import Dependence

Before Lithuania re-established its independence, its power system was an integrated part of the Soviet grid, and although oil, natural gas, coal and uranium were imported, the country was a net exporter of electricity. The total installed capacity of Lithuania's power plants is 6,324 MW, including the Ignalina nuclear power plant (NPP), with 3,000 MW of installed capacity (later reduced to 2,500 MW for safety reasons) and the Lithuanian thermal power plant (TPP) with 1,800 MW of installed capacity, along with significant combined heat and power plant (CHP) installations, and some hydropower.

Until 1991, significant amounts of electricity generated in Lithuania were exported to Belarus, Latvia and the Kaliningrad region in Russia. Lithuania and its neighbours are currently in a deep economic crisis, however, which has reduced electricity demand both internally and in neigbouring countries.

The Ignalina NPP has two reactors, each with a capacity of 1,500 MW. The reactors are of the RBMK type (the same model as at Chernobyl). The first reactor was commissioned, and the station started to produce electricity, in 1984. The second reactor was put into operation in 1987. Ignalina generates about 85 per cent of Lithuania's total electricity demand and operates about 3,940 hours per year (which gives it a so-called 'average annual load' of 45 per cent).

The first unit at the Lithuanian thermal power plant was commissioned in 1963 and total installed capacity is now 1,800 MW (four 300-MW capacity units and four 150-MW capacity units). All 300-MW units are condensing units (used for electricity production only), and two of the 150-MW units also supply heat to the surrounding area. Two 150-MW units of the Lithuanian TPP have been refurbished, and further modernization is planned, based on the results of a recent pre-investment feasibility study. Despite its huge potential, the TPP is in a 'cold regime' and operates only about 306 hours per year. This gives it an 'average annual load' of only 3.5 per cent. In fact the plant is only kept in existence as a back-up to replace the Ignalina plant in the case of an emergency. This, of course, puts a huge financial burden on the state economy.

There are three large, modern combined heat and power (CHP) plants in Lithuania: in Vilnius, Kaunas and Mazeikiai. There are also several small, older public CHPs and industrial co-generation plants. Due to the same problems affecting Lithuania's TPP, the average annual load of CHPs is only about 11.5 per cent.

Industrial co-generation plants are located within industries where there is a demand for process steam. The plants are in mineral fertilizer factories in Joanna (24-MW capacity) and Kedainiai (10-MW), and in paper-mills at Grigiskes (5-MW) and Klaipeda (12-MW). The total installed capacity of these plants is 51 MW.

Lithuania's only large hydroelectric power plant is at Kaunas on the Nemunas river, with an installed capacity of 100.8 MW. The plant was constructed in 1960, and replacement of turbine-runners, generator windings and other equipment would increase its lifetime by 30 years. The average annual load of the plant is just over 40 per cent.

Lithuania is almost entirely dependent on energy imports, and indigenous energy resources (hydro, peat and fuelwood) supply only around 5 per cent of energy demand. There is a potential to improve the use of indigenous energy and also some potential for the use of renewable energy, especially in the case of geothermal, hydropower and biomass. Specifically, opportunities exist for small hydropower rehabilitation, and conversion of heat-only systems to CHP using wood waste and wood chips.[2] The country also has some other indigenous energy resources, including onshore and offshore oil, peat deposits and natural gas, but these amounts are either insignificant, or their exploitation would be likely to have negative environmental consequences.

Lithuania's hydropower output averages 320 gigawatt hours (GWh) per year, and it makes up only 10 per cent of technically harnessable hydropower resources. Only Lithuania's two biggest rivers, the Neris and Nemunas, have total potential capacities of more than 100 MW. Medium and small rivers are estimated as having about 120 MW of potential capacity. Total hydropower potential is estimated at 6,000 GWh, of which 3,600 GWh per year is technically feasible.

There is a large potential for small and micro hydropower plants, from which about 500 GWh could be produced annually.[3] Unfortunately, the Lithuanian National Energy Strategy considers the economic viability of new hydro capacity to be economically restricted, and rehabilitation of only 8 MW of small hydropower capacity has been planned by 2000.[4] Some private companies have recently approached the EBRD for loans to rehabilitate 13 of the country's small hydropower plants.

Lithuania has a strong potential for the use of biomass fuels. Potentially, 52 per cent of Lithuania's electricity could be produced from fuelwood and as much as 23 per cent could be produced from cereal crop residues. These biomass fuels could be economically competitive with both coal and oil, moreover, even without considering their environmental advantages.

The technical potential of geothermal energy within Lithuania's borders corresponds to 6.8 billion tons of oil equivalent (TOE), or an economic asset value of approximately US$640 billion.

Investigations and calculations of the National Energy Efficiency Programme have shown that approximately one-quarter of currently consumed energy supplies could be saved over the next ten to fifteen years by the introduction throughout the country of energy-saving measures expected to be realized in the programme and in the National Energy Strategy. According to these calculations, the potential for energy savings represents 30 per cent of all primary energy consumed. Insulation of buildings could save about ten terrawatt hours (TWh) of this potential, but it demands relatively large investments. Other measures, including implementation of energy accounting and improvements in management, could be introduced with comparatively small expenditures to produce very good economic returns. In general, not enough energy-saving measures have been introduced to date, and not enough attention is being paid to this field in Lithuania.

The Bank's Focus on Energy Supply

The World Bank's energy-related projects in Lithuania have concentrated on the supply side. Relatively little attention has been paid to the Bank's stated priorities of promoting demand-side energy efficiency. A loan for an Energy Efficiency/Housing Pilot Project is the only demand-side oriented project supported by the Bank to date, and even this has faced implementation problems caused by the regulations and high interest rates of local commercial banks.

The Bank's principle of encouraging private sector investments in the power sector is also not well developed in Lithuania. It has so far focused only on regulatory and legislative issues, giving no financial support to facilitate the involvement of independent private investors in the promotion of new and renewable energy technologies.

The World Bank is involved in: a US$26.4 million Power Rehabilitation Loan intended to improve operating safety, efficiency, reliability, and environmental performance of the thermal electricity generating system, thus reducing the amount of imported fuels needed for its operation and facilitating the retirement of the Ignalina plant; and a US$5.9 million loan for the Klaipeda Geothermal Demonstration Project to demonstrate the feasibility of developing indigenous Lithuanian geothermal energy resources. The total cost of the project is just over US$18 million and other investors are the GEF, the European Union grant programme for Central and East European countries (PHARE), and the Danish Ministry of Energy. In theory, the project has positive long-term consequences. Its implementation has been heavily criticized, however, by specialists from the Lithuanian Energy Institute, and others, for its inadequate analysis of geothermal resources and lack of clear benefits. A US$10 million loan is intended to increase demand-side energy efficiency in residential and public buildings and to support the implementation of governmental policies on the privatization of housing, enabling increased private initiative in housing maintenance. This is the only demand-side project in Lithuania to date, and it has run into problems due mainly to the internal regulations and the high interest rates charged by local commercial banks. A US$80 million structural adjustment loan has been to restructure Lithuania's economy and energy sector along commercial lines. The Bank also has a project under preparation to modify network connections in the Klaipeda District Heating System. This would comprise the replacement of obsolete equipment, including the installation of new thermostats and circulation pumps. It would also provide for the establishment of a workshop for the assembly of sub-stations, and supervision of equipment installation.

The stated energy policy of the EBRD aims to assist countries to reorient their energy sector development towards more efficient and 'least cost' options, and to give priority within this to supporting the private sector. Nevertheless, the EBRD has not yet supported private sector involvement in new and renewable energy.

Specifically, the EBRD has provided loans for a US$46.25 million, made in 1992, for technical assistance to begin the process of commercial reorientation of Lithuania's energy sector, the 'Necessary Investments in the Energy Sector' project. Although the project was designed for the improvement of energy efficiency, commercialization of the energy sector, improvement of environmental performance, and related issues, the actual use to which the loan has been put is far from clear. The Lithuanian Energy Agency has not been able to give any examples of energy efficiency measures that have been financed under the project. Nor are there any examples of measures for environmental performance that have been supported under the project. About 5 per cent of the loan was used for investments to improve energy sector performance, while the rest was used for purchasing fuel and other operational and consumption costs;[5] a grant of US$ 42 million from the Nuclear Safety Account (NSA) for urgently needed safety upgrades to the Ignalina Nuclear Power Plant. The NSA was established in 1993 at the request of the G7. Thirteen donor countries and the EC fund it, and it is administered by the EBRD. Prepared in close cooperation with the G24 Secretariat[6] and bilateral assistance programmes, in particular with Sweden, the project is helping to implement a Safety Improvement Programme by providing equipment for operational and technical safety improvements and by funding project management and engineering experts. Official statements in the Lithuanian media have been promoting the myth of cheap nuclear energy and claiming that the Ignalina NPP is necessary and important for the Lithuanian economy. In the National Energy Strategy, operation of the nuclear power plant is closely connected with the country's economic development and it is very clear that if the export of electricity becomes possible for Lithuania, operation of both Ignalina units will be extended for as long as possible. Recently, the government has been looking for possibilities to join the Western European Electricity Network (UCPTE), and has very actively promoted the export of electricity. During several rounds of high-level negotiations on joining UCPTE, various possibilities were discussed, including a scheme through the Nordic countries – laying cable under the Baltic Sea – and the laying of a cable through Poland. Eventually, at the beginning of 1998, the Lithuanian Ministry of Economy proclaimed an international tender for construction of a 110-kV electricity transmission line to Poland. According to the tender's conditions, the winning bidder, in addition to paying for construction of the transmission line, will have to purchase 6 TWh of electricity annually during a ten-year period and electricity export is anticipated to start in 2002. This means that the Lithuanian government is now planning to operate both units at the Ignalina NPP until at least 2012.

Accelerating Towards the Market

The objective of the World Bank, and other MDBs, in Lithuania is to accelerate the country's further economic transformation towards a market system, which the World Bank, for example, says will produce a recovery of living standards and output and export growth. To reach this objective, the World Bank mainly assists in the implementation of set government policies. In the energy sector, the Bank provides the government with sectoral research, provides project loans, and manages the country's structural adjustment loan. The Bank's policy advice provides an analytical review of the energy sector as well as needs and possibilities for future lending, but it is mainly oriented towards the further development of oil, coal, gas, nuclear, and supply-side management.

Recommendations concerning the development of the Lithuanian energy sector and corresponding national energy policy have been closely related to World Bank studies. The Bank has not encouraged the government to look for non-nuclear alternatives to energy generation. The Bank's position has been simply to investigate different scenarios set out in the National Energy Strategy and to find the most 'beneficial' means of power generation. According to a study of Lithuania's energy sector carried out for the G7, called *Lithuania: Power Demand and Supply Options*, the 'high nuclear' scenario is the most efficient in economic terms, as the option demanding the lowest investments. This 'high nuclear' scenario would allow both units at Ignalina to operate to the end of their useful economic lives.[7] However, as the most feasible option, the Bank proposes to retire one of Ignalina's units earlier, and the second unit by 2000.[8]

In the Bank's policy advice to Lithuania, little attention is paid to energy efficiency and the development of renewable sources of energy. The only general recommendations on energy sector development concentrate on rehabilitating existing facilities and retiring, in a planned manner, those that are unneeded, inefficient, obsolete or unsafe. Only two bank projects address the development of renewable energy sources, energy efficiency and demand-side management. These are the Energy Efficiency/Housing Project and the Klaipeda Geothermal Demonstration Project, both described above, which make up only 6.3 per cent of all MDB energy investments in Lithuania.[9]

Notes

1. This analysis was prepared by Saulius Piksrys, Lithuanian Green Movement, Kaunas, Lithuania.

2. *The European Renewable Energy Study II (TERES II)*, Consultation Report Lithuania, Altener programme, 1995, p. 7.

3. Ibid., p. 18.

4. *National Energy Strategy*, IC Consult, Erm Energy Ltd, Cowi Consult, Lithuanian Energy Institute, 1993, p. 118.

5. Personal communications, Lithuanian Energy Institute, 1997.

6. The G24 is the Intergovernmental Group of Twenty-Four on International Monetary Affairs. It was established in 1971 with the aim of promoting the position of the developing countries on monetary and finance issues.

7. *Lithuania: Energy Sector Review*, World Bank, 1994, p. 79 (Table 7.3).

8. *Lithuania: Power Demand and Supply Options*, World Bank, 1993, p. 55.

9. Economic newspaper *Litas*, 'Foreign loans of the Republic of Lithuania as of 1 October 1996'.

13

Ukraine[1]

The Highest Energy Intensity in the World

The energy industry in Ukraine includes the coal, oil, gas, peat and refinery sectors. Ukraine has large coal reserves, as well as oil, gas and other fossil fuel resources, but their extraction is not currently sufficient to meet domestic demand and their share of domestic primary energy products is relatively low.

Ukraine's official energy policy has the following stated aims: the development and implementation of a policy that promotes energy conservation; economically and environmentally justified utilization of domestic energy sources; restructuring of the economy to reduce energy intensity; and increasing reliance on alternative (renewable) energy sources.[2] Meanwhile, Ukraine's priorities for the power sector are: extensive rehabilitation of existing thermal power plants; promotion of new technologies for the clean burning of low-quality coal; utilization of efficient gas turbine equipment; building of new nuclear units and reconstruction of already existing ones; capacity-building in nuclear waste management; completion of new hydro-power plants and pump storage plants and utilization of small- and medium-size rivers for power generation; and rehabilitation of existing hydropower plants.[3]

Energy intensity in the Ukrainian economy is the highest in the world and three to four times higher than most industrialized countries, mainly due to the large amount of heavy industry and poor energy efficiency in the country. The National Energy Programme predicts, however, that it is possible to reduce energy consumption by 10 per cent – an amount equivalent to 30 million tonnes of coal – without additional expenses or with only small investments. Further, with extensive controls for energy resource utilization, new pricing policies and additional investments, energy savings could be 1.5–2 times higher than that. The National Energy Programme also indicates a potential for the development of renewable energy in the country, particularly solar, wind, geothermal, small hydro, wave power, and the utilization of biogas and mine methane.

Sustainable Options Not Fully Considered

The World Bank has been involved in a US$114 million loan for the rehabilitation programme of the Kakhovka hydropower station; and partial implementation of the rehabilitation programme for the Kiev, Kanev, Kremenchug and Dniprodzerzhinsk hydropower plants, installation of dam safety monitoring systems, upgrading of communications, dispatch, system and protection, and generating unit controls and assistance for project implementation, and optimization of use of the reservoirs on the Dnieper river; and a US$300 million Coal Sector Adjustment Loan to restructure the coal sector, including corporatization, price-, trade-, and export-liberalization, decommissioning of unprofitable mines and investment in profitable corporatized mines. The loan also includes a certain amount for 'social mitigation'; a US$317 million loan for the Electricity Market Development Project, which includes building up fuel stocks at 14 thermal power plants to levels that are consistent with standard industry practices (stocks for about 40 days), building up stocks of spare parts and carrying out deferred maintenance at the plants. It also includes installation of metering and communications equipment to improve recording and building of electricity flows at key wholesale market delivery points, and technical services and training for project implementation, financial management, and the development of a privatization programme. World Bank projects in the pipeline at the time of writing are: a US$100 million loan for gas distribution rehabilitation, which will support the rehabilitation and upgrading of the gas distribution system in four areas, installation of gas meters for customers in the residential/services sector, rehabilitation and/or replacing sections of the pipeline networks, and an institutional support programme; a US$260 million loan for the Dniester Hydropower Pump Storage Project, which aims to complete three units of the Dniester Hydropower Pump Storage Plant (DHPSP), completion of the after-bay (buffer) hydropower plant at the lower dam of the DHPSP, strengthening of the transmission system, further priority upgrades of the dispatch control and communications systems, and technical assistance for project implementation; and a US$200 million loan for Kiev District Heating Improvement. The project includes the rehabilitation of and introduction of technologies and materials to the heating system in Kiev and support for the commercialization and strengthening of project district heating companies; and US$160 million for the Krivoy Rog Power Plant Rehabilitation Project to facilitate general station rehabilitation at a major coal-fired thermal power plant to extend service life, increase efficiency and reduce environmental impact, including boiler and turbine-generator rehabilitation.

European Bank for Reconstruction and Development (EBRD) projects in Ukraine include: US$61.9 million for the Power Market Development Project for short-term repairs and maintenance at the main thermal power stations and to provide new meters and equipment for the new electricity wholesale market; US$8 million for the Poltava Oil and Gas Extraction Project for the drilling of four new wells, connecting these wells to the operation and production base, and constructing a pipeline and rail export facilities; and US$113.22 million for the Starobeshevo Power Modernization Project to replace an old coal-fired boiler with the installation of a 210-MW – more energy-efficient – fluidized bed boiler with ancillary equipment at the Starobeshevo Power station in Eastern Ukraine. EBRD projects in the pipeline at the time of writing include US$62 million for the Lviv District Heating Commercialization Project to commercialize the district heating supply service in Lviv. Investments will assist the municipality in developing its utilities on a commercially and financially autonomous basis, enabling private agents to handle customer services, metering, billing and collection, with US$124 million for Kiev District Heating Rehabilitation to finance the rehabilitation of Kiev's transport and distribution district heating grid; US$290 million for the Krivoy Rog Power Plant Rehabilitation, which consists of rehabilitating three units at Krivoy Rog, improving their efficiency, and reducing emissions of air pollutants; and US$33 million for the Ukrainian Energy Service Company (UkrESCO) Project to establish the first Energy-Service Company (ESCO) in Ukraine and implement a range of energy-saving projects in the public and private sectors.

Electricity Market Rehabilitation Project

In 1995, the government of Ukraine agreed with the EBRD and the World Bank to undertake the Electricity Market Development Project. The EBRD's contribution – totalling US$61.9 million – is now being implemented. In 1997, the World Bank suspended credits for its US$317 million contribution, however, after the Ukrainian government refused to raise sharply energy prices for residential consumers, in order to reflect the higher cost of residential energy delivery, as was specified in the loan agreement.

The objective of the project is to support the development of a competitive electricity market by providing working capital to thermal power generators for the purchase of fuel and spare parts, fuel quality improvement, and the installation of metering and communication equipment needed for proper functioning of the system.

Although the project includes components to improve both demand- and supply-side energy efficiency, these are only indirect results of its main objectives. In fact, efficiency measures involve no more than switching from low- to higher-quality coal. No proposals for coal-oil, coal-natural gas, coal/mazut-natural gas have been presented, and it would appear that these options – along with options for demand-side efficiency – were not taken into consideration during the project's development.

Heat Supply and Energy Efficiency Project

A US$200 million World Bank loan was provided for this project, with the aim of rehabilitating existing heat-generating capacity. One of the two beneficiaries for the project is Kievenergo, which owns one of the biggest heating systems in the Ukraine district and operates both heat-generating and combined heat and power units. As such, the project could have been a good demonstration or pilot project for the promotion of co-generation technology in Ukraine. Unfortunately, this opportunity was missed.

A 1996 report on Ukraine's energy policy by the International Energy Agency (IEA) discovered that Kievenergo supplies only 5 per cent of power consumption in Kiev, because initial investments went to the development of heat generation, rather than co-generation.[4] The IEA report came to the conclusion that the future effectiveness of the heating system in Kiev depends on the modernization of coal-fired boilers, coupled with a switch to co-generation with boilers using natural gas.

The World Bank-funded project ignores the IEA's findings, however: it includes no proposal for switching to co-generation, and is intended, rather, to prolong the lifetime of existing boilers, thus creating an additional barrier to future investment in co-generation in Ukraine.

The World Bank has been involved in the Ukrainian energy sector since early 1992. The Bank completed an Energy Sector Review in 1993,[5] and this led to an Energy Strategy Conference, which was held in Kiev in June 1993. The Bank made it clear at this conference that it would focus on power generation, gas transmission and gas distribution. It was also determined that lending operations should focus on the rehabilitation of existing assets rather than on capacity expansion, while supporting initiatives that increase the financial and operational autonomy of enterprises and foster competition. The Bank's most important investment in this direction has been the Electricity Market Rehabilitation Project, described above, which pays little attention to demand-side efficiency.

At first glance, it might appear that the World Bank is intending to give more attention to demand-side efficiency in its future investments in Ukraine. The Bank's 1996 *Electricity Market Development Report*, for example, states that: 'An analysis of the Ukrainian power system development for the period 1996–2010 has been performed to assess future investment strategies to maintain the system's ability to supply expected electricity demand. The analysis was performed using the least-cost methodology.'[6]

The problem with the above analysis, however, is that it did not take account of all available energy options. In fact demand-side efficiency and renewable energy options were not examined at all. Rather, the Bank's 'least-cost' analysis examined thermal generation options, pump storage plants and three nuclear power units that are under construction. The consequence has been, of course, that the Bank's *Market Development Report* only recommended supply-side investments for the future of Ukraine's energy sector. This is in direct contravention of the Bank's 1992 energy policies, which require that: 'World Bank lending in the energy sector should be based on and, where necessary, support as part of country assistance strategies the development of integrated energy strategies that help borrowing countries take advantage of all energy supply options, including cost effective conservation-based supplies and renewable energy sources.'[7]

The objectives of the EBRD in Ukraine are to help reduce the economy's energy intensity, facilitate the closure of Chernobyl, and improve overall environmental performance in the energy sector. Specifically, the EBRD's priorities are: to improve efficiency and environmental performance of power generation and in commercially structured projects sponsored by private investors; promote improved end-use efficiency, and more rational use of energy through price reform and energy efficiency initiatives; help to improve the reliability of the power and gas transport systems; support reorganization of the sector; and assist in the development of domestic fossil fuel reserves. This strategy is directly linked to the G7 Action Plan for Ukraine, which proposed a comprehensive package of nuclear safety upgrades, tariff increases, rehabilitation of thermal plants and completion of modern nuclear plants under construction to facilitate the earliest feasible permanent closure of Chernobyl.[8]

Lahmeyer International performed a least-cost analysis of the Ukrainian power sector for the EBRD in July 1995, but the EBRD strategy for Ukraine for 1995–96 had already been approved in February 1995. The findings of the analysis coincided with the Bank's major strategies for Ukraine, and the EBRD's proposed projects generally reflect this.

The EBRD's commitment to least-cost planning has been tested by two incomplete nuclear power plants at Khmelnitsky and Rivne. The EBRD had agreed to go ahead with the completion of these nuclear power projects only if they were clearly the least-cost options. A panel of independent experts, commissioned by the EBRD itself, concluded in February 1997, however, that these projects – known collectively as K2/R4 – would not be the best option. The panel's statement made this very clear, and said:

'We conclude that K2/R4 are not economic. Completing these reactors would not represent the most productive use of US$ one billion or more of EBRD/EU funds at this time.'[9] Nevertheless, at the time of writing, the EBRD is still procrastinating and has not made the decision to suspend the Rivne and Khmelnitsky projects.

The independent panel mentioned above also recommended using the huge energy efficiency and energy conservation potential in Ukraine for solving its current energy problems, but these recommendations have also had little impact in terms of actual energy investments in Ukraine.

The lack of attention being paid by the WB and EBRD to demand-side energy efficiency in Ukraine is also clear from a report carried out by the International Institute for Energy Conservation (IIEC) for the OECD Environment Directorate.[10] The report includes a detailed analysis of approved and proposed projects to be funded by the WB and EBRD, which address various aspects of energy efficiency. In short, the report concluded that none of the six World Bank- and EBRD-approved projects can be classified as a completely demand-side energy efficiency project, only two approved projects include secondary measures to increase supply-side energy efficiency, and only one project includes secondary energy-efficiency components for both the demand and the supply side.

The report's overview of the projects in the pipeline, however, gives a more optimistic picture: six of the projects presented invest directly in energy-efficiency measures, four of the projects address demand-side efficiency, and two projects address both demand- and supply-side efficiencies. Four of these are EBRD projects, and two are World Bank projects. The total loan amount is about US$510 million. Nevertheless, many projects in the World Bank's or EBRD's pipeline, such as the Dniester Pump Storage Project (Ukraine Hydropower–II), and the Coal Restructuring Project, still include no apparent energy-efficiency components at all.

Moreover, although the National Energy Programme of Ukraine includes special plans for the development of new and renewable sources of energy such as wind, solar and geothermal energy, coal methane utilization, and biogas production, the EBRD and World Bank have yet to consider even one such project for Ukraine.[11]

Notes

1. This analysis was prepared by Yurij Urbanskyi, National Ecological Centre, Uritskogo 15–115, Kiev 252035, Ukraine. A full-length report is available in Ukrainian. A more detailed analysis appears in *Time for a Change*, which can be downloaded via www.bothends/Mdb

2. *Concept for the Development of the Energy Sector of Ukraine for the Period up to 2010*, Government of Ukraine, Kiev, November 1993, p. 21.

3. *The National Energy Programme of Ukraine up to 2010*, May 1996, p. 34.

4. *The Energy Policy of Ukraine, the 1996 Survey*, International Energy Agency, p. 119.

5. Report no. 11646.

6. *Electricity Market Development Report*, World Bank, 1996, p. 1.

7. *Energy Efficiency and Conservation in the Developing World*, World Bank, 1992.

8. *EBRD, Strategy for Ukraine for 1994–1996*, European Bank for Reconstruction and Development, 1994, p. 4.

9. *Economic Assessment of the Khmelnitsky 2 and Rovno 4 Nuclear Reactors in Ukraine, Report to the European Bank for Reconstruction and Development, the European Commission and the US Agency for International Development*, International Panel of Experts, chaired by Professor John Surrey, Science Policy Research Unit, University of Sussex, 4 February 1997, p. 68.

10. *Financing Energy Efficiency in Countries with Economies in Transition, A Study for the OECD Environment Directorate*, International Institute for Energy Conservation (IIEC–Europe), OECD, 1996.

11. *National Energy Programme of Ukraine*, May 1996, pp. 167–79.

14

Brazil[1]

Energy in Transition

Traditionally, the Brazilian energy sector has been characterized by the prevalence of two major state-owned holding companies, Eletrobrás and Petrobrás, the state ownership of the electric utility network, and state planning. But with radical reforms under way, this structure is rapidly changing.

Two agencies are in charge of the two main energy sub-sectors, electricity and hydrocarbons. The National Electricity Agency, ANEEL, sets tariffs either directly through authorization, or indirectly through concessions or licences. Eletrobrás controls four regional companies, two distributors, holds minority shares in the companies of several states and coordinates the planning for expanding the operations of electricity systems. Eletrobrás is also involved in financing and channelling funds for the development of the sub-sector. But the number of players in the Brazilian energy sector is increasing as publicly owned state electricity utilities are being broken down into generation and distribution companies, and privatized. New entities are also appearing with the opening of the hydrocarbons sector to private participation and the construction of large gas pipeline networks.

In 1997 the National Oil Agency, ANP, established a framework for deregulation of the hydrocarbons sector grants, concessions for exploration and exploitation of oil and natural gas to Petrobrás, private companies or joint ventures. Although no longer a monopoly, Petrobrás participates in all stages of oil and gas production and marketing, pursuing a strategy of building broad alliances with the private sector. The National Bank for Economic and Social Development, or BNDES, invests in energy sector enterprises and sometimes finances the activities of transnational corporations that have been awarded privatization contracts in the energy sector, such as the buyers of the power distributors Metropolitana and Bandeirantes of São Paulo.

Between 1980 and 1995, total energy consumption in Brazil grew by 43 per cent, much of which was met by a doubling in the use of hydropower. In 1991, 74.2 per cent of households were electrified, but 4.35 million houses were without electricity and another 4.84 million were connected to the grid illegally, with the implied safety risks. Despite its dynamic growth, most of Brazil's hydroelectric potential remains unexploited. Oil and its derivatives provide Brazil's second energy source, but are in decline. Between 1980 and 1995, Brazil reduced its dependence on imported oil from 80 per cent to 43 per cent by increasing refining capacity and oil production, mainly from off-shore exploitation in Rio de Janeiro. Natural gas provides a small proportion of Brazilian energy at just over 2 per cent, but consumption is increasing more rapidly than any other source and is expected to make up 10 per cent of the energy mix in a few years. There are several projects, some already under way, for pipelines to carry Bolivian and Argentine natural gas to Brazil, the idea being to use natural gas for thermoelectric generation. The low price of fuel oil, however, poses severe limitations on a natural gas substitution policy. Further, fuelwood accounts for 12 per cent of Brazil's energy supply. The contribution of fuelwood is dropping mainly because of a decrease in residential consumption. Sugarcane became an important energy source through the Alcohol National Programme (PROALCOOL), launched in 1975 to reduce dependence on oil imports for transport. US$4 billion of public funds were invested to increase ethyl alcohol production, which is currently around 11,000–12,000 million litres per year. Most farmers in São Paulo substituted sugarcane for other crops, but when the international price of sugar rises, producers turn to foreign markets, which, in turn, leads to import of sugarcane to avoid shortages in fuel production. Nearly 33 per cent of Brazilian vehicles (around 4.5 million) run on alcohol. The use of non-conventional energy sources (solar, wind, biomass) is not significant. The Brazilian nuclear programme has not yet resulted in power generation.

Regulation of the electricity and hydrocarbons sectors is dramatically changing in Brazil. The model in the electricity sector used to be that of regional monopoly with captive markets, vertically integrated companies, federal and state centralization and a cooperative process among the sector's actors. The traits of the new model are competitive generation and distribution, and the separation of transmission and distribution functions into independent enterprises, private capital participation (with the subsequent increase of actors in the sector), and the coexistence of captive and free markets. The general guidelines for the new electricity model as established by law are:

- competition in electricity generation;
- free access by any agent to transmission grids or transportation networks;

- freedom for large consumers to choose their power supplier;
- the creation of a new legal entity, the independent power producer (IPP); and
- the need for one simple 'agreement' between transmission actors.

The ongoing privatization process of regional and state enterprises and of those of the various Brazilian states covers only generation and distribution, while transmission remains public property. The new framework aims at creating competitive conditions in the power sector through the participation of an increasing number of private agents. The IPPs are expected to transform the sector by exerting legitimate pressure on the present monopolies.

Regulation functions, including decisions about concessions for exploration and exploitation of oil basins, were delegated to the National Oil Agency. Petrobrás was not privatized because of its significance in the national economy; its operations amount to nearly 2 per cent of Brazilian GDP. If it operated as a conglomerate and purchaser of goods and services in the domestic market, this figure would rise to 25 per cent.

The new legal framework entitles Petrobrás to undertake joint ventures with the private sector, either as a majority or minority partner, and to sell assets of such associations without previous consent from the Brazilian Congress. Foreign companies seek association with Petrobrás because of its 40 years of experience as the only oil and gas explorer and producer in Brazil, and because it is a recognized leader in the deep-water technology necessary for the exploitation of marine crude oil reserves. Petrobrás plans to invest US$5 billion in oil and derivatives exploitation and distribution, in association with 37 foreign companies, including Exxon, Shell, Mobil, British Hydrocarbons, Amerada Hess, Elf and YPF.

Evolution of Brazilian energy production and consumption over the last 15 years has not been the result of an integrated energy policy. Rather, a series of sectoral plans and programmes were developed with the private sector in isolation from the public. One such sectoral programme was the PROALCOOL Programme for replacing fossil fuels, which absorbed huge state resources and accumulated massive bad debts from the programme's target producers. Present energy policy aims to:

- increase power generation capacity using the abundant hydroelectric resources available and progressively intensifying the conventional thermal power generation from natural gas or coal, interconnecting the north/north-east and south/south-east/midwest electricity systems, and establishing a Brazil–Argentina interconnection for power importation;
- incorporate natural gas (from Bolivia and Argentina) into the energy matrix;

- develop the existing mineral coal reserves in the south of Brazil;
- develop the capacity of oil production within the country and abroad; and
- slow the pace of the nuclear programme.

Brazil's neighbouring countries, Venezuela, Argentina, Peru and Bolivia, all have large natural gas reserves, but in Brazil gas provides barely 2.5 per cent of energy consumed. Brazil's decision to increase gas imports is expected to increase this figure to 12 per cent. Several pipelines have been designed to carry natural gas into Brazil and once the necessary thermal plants have been installed, gas will provide electricity to the industries of São Paulo, Porto Alegre, Florianópolis and Curitiba. Natural gas is being promoted as a reliable abundant fuel on the grounds that it has environmental advantages, but its main attraction is its comparatively low cost. There are, however, still short-term barriers to increasing the role of gas in Brazil. Delays have occurred in the construction of gas-fired thermoelectric plants and industries are just starting to convert to natural gas-burning equipment. Very few of the 20 projects for gas-fuelled thermal plants, eleven of which lie along the Bolivia–Brazil pipeline, have actually been implemented. Uncertainty about the profitability of thermoelectric plants seems to be the cause of delays. This is partly because during the rainy season hydroelectric power is cheaper than gas-generated power.

A Chance for Sustainable Energy?

The National Programme for Electricity Conservation, PROCEL, was launched following electricity conservation studies that led to a joint action of the former Ministries of Mining and Energy and of Industry and Trade, which established PROCEL, the first systematic effort to promote the rational use of electric power. PROCEL has improved the analysis of basic market characteristics and identified wasteful tendencies in the electricity sector. It develops projects to promote waste reduction and improve equipment and systems to use electricity more efficiently by employing institutional, financial, managerial and promotional mechanisms to reduce electricity consumption in each sector. One project, for example, identified major energy losses and opportunities for improvement in several companies and assessed their energy conservation potential. Simple and inexpensive waste elimination measures are presented, which may involve resizing engines, equipment or adjusting production processes, electric installation upgrading, or improving management. To produce measurable effects, PROCEL sets efficiency targets per end use, for example, for lighting, refrigeration, air-conditioning, ovens and engine systems as well as for sub-sectors such as industry, residences, services and public lighting, based on international and national assessments of electricity consumption and penetration of new technologies. These targets, amounting to a consumption reduction of 12 per cent within 25 years, are incorporated in long-term planning. By adding the small gains obtained in the electric system and from consumers, conservation will play a significant role in the power sector. Encouragingly, plans have been made to introduce energy conservation as a normal premise of engineering so that it ceases to be an auxiliary activity detached from planning, design, implementation and operation of policy. PROCEL has obtained direct and measurable economies above 1.2 terawatt hours (TWh), with costs below US$6 per barrel of oil equivalent, at costs of less than 20 per cent of the cost of power system expansion. It has likewise accomplished some indirect and induced economies.

In Brazil, non-conventional energy sources such as solar, wind and biomass present an alternative for isolated communities or low-income areas throughout the country. Their contribution to energy supply, however, is still low. Studies have shown that, depending on factors such as distance from the grid, number of residences to be served, and load to be supplied, generation from photovoltaic cells seems to be more cost-effective than grid extension. But the profitability of such investment is uncertain given the poor conditions of target populations. Therefore, photovoltaic dissemination to this target group can be accomplished only through charity from the public utilities and international assistance. It is estimated that Brazil can develop a power capacity of 50 MW in photovoltaic generation and three million square meters of thermo-solar capture.

There are two outstanding examples of wind generators: one 75-kW generator installed by CELPE/UFPE in Fernando de Noronha and a set of four 250-kW generators connected to the network, installed by CEMIG on Carmelinho hill in the state of Minas Gerais. Brazil's potential for wind power is estimated at 1,000 MW. The electric utility Idaho Power Company, with 17 hydroelectric and four thermoelectric units in the state of Idaho (USA), will open a Brazilian subsidiary to install mixed solar–wind power facilities in isolated places that are currently serviced by thermoelectric units based on gas oil. The company relies on the latest technology in renewable energy generation systems for localities far from the distribution networks.

The 1973 oil crisis fostered studies of biomass energy resources, including ethanol made from amylaceous and cellulose materials, methanol from wood, vegetable oils, and organic and industrial wastes. Despite their high estimated potential, the projects did not attract investors and were dismissed because of their low profitability after the 'oil shock'. Primary power generation from biomass occurs at the Biomass Integrated Gasifier-

Gas Turbine developed by a consortium including Chesf, Eletrobrás, Cientec, Vale do Rio Doce, Shell and the Ministry of Science and Technology, with World Bank assistance.

The Action Plan for the Development of Renewable Solar, Wind and Biomass Energy in Brazil, designed by the Renewable Energies Forum, set the following goals for 2005:

- 3,000-MW power capacity in co-generation from sugarcane waste pulp;
- 1,000-MW power capacity in co-generation from wastes of the paper industry;
- 250-MW power capacity in thermoelectric units running on fuelwood from plantations;
- 150-MW power capacity in small-scale generation systems fuelled by vegetable oils;
- 12 million tons charcoal/year; being the whole growth referenced to present production (around 10 million tons/year) obtained by sustainable means;
- 18,000 million litres/year of ethyl alcohol for fuel purposes;
- 20 million litres/year of fuel vegetable oils;
- 80,000-m^3 biogas obtained from urban, industrial and rural wastes; and
- 3 million additional hectares reforested with native and exotic species.

The Banks and Market Restructuring

Lack of investment in transport and the energy infrastructure in recent years has increased the cost of production and jeopardized the success of Brazil's industrial modernization strategy. For this reason, the World Bank proposed a joint analysis of the various alternatives to the Brazilian authorities, to ensure the financial feasibility not only of individual projects but also of global sectors. The Bank is offering to assist reforms that improve the institutional, legal and political structures in the productive sectors, by strengthening tariff and regulatory legislation and setting transparent rules to ensure autonomy for private sector operations in both the physical and energy infrastructures. Examples are adjustment of power tariffs to real costs, creation of less monopolistic conditions, and decentralization. In 1995 approximately eighteen power projects were stalled at different stages of planning and construction because funds were lacking, according to the *Brazil Country Paper* published by the Inter-American Development Bank in 1995. This jeopardizes the government strategy for modernization of the productive sector, including the feasibility of energy rationalization. On account of the above, the World Bank strategy in the electricity sub-sector also includes:[2]

- institutional strengthening with the creation of new agencies and reform of existing ones, to improve conditions vital to long-term planning, such as engineering, construction and operation of complex systems, etc.;
- improvement of public services, better policies for staff and financial management, and development of better technologies for energy generation, transportation and marketing; and
- projects that contribute to priorities, such as energy conservation, loss reduction, diversification of the energy matrix, privatization, stronger private sector involvement in resource mobilization and operation of concessions, and efforts to reduce (import) prices through financial resources.

At present, 37 projects are being implemented (and 22 are projected for future consideration) with World Bank assistance, with investments totalling US$6.1 billion for the period 1995–97. Of this amount, only US$30.5 million was earmarked for the energy sector, for projects such as the Brazil–Bolivia pipeline; a pilot biomass project in north-eastern Brazil; an energy efficiency project; a hydrocarbon transport project; a gas sector development project; and the Segredo Hydroelectric Project of Compañía Paranaense de Energía (CEPEL). In addition, the World Bank also offers assistance to several power utilities and some small to medium-size industries through loans from the state-owned National Bank for Economic and Social Development (BNDES).

Brazil's most significant economic restructuring is taking place in the power sector. The generation and distribution of energy are being privatized, while transmission is being kept as a natural monopoly. Some of the companies targeted for privatization are very profitable, such as Furnas, which earned US$360 million in 1997. According to Dennis Jungerman, vice-president of the mergers and acquisitions division at JP Morgan, a major advantage of privatizing Brazilian generators is their high efficiency level. 'In terms of operation, organization and staff technical quality, generation plants are in excellent condition. The problem with some of them is that they are heavily indebted' (*El Cronista*, 31 August). BNDES finances 50 per cent of electric sector privatization with foreign resources. This means that the state borrows money in order to loan it to the buyers, which are mainly large multinational corporations, thus increasing the country's already high foreign debt, at least in the short term. Privatization of São Paulo's two distribution utilities, Metropolitana and Bandeirantes, required a loan of approximately US$1.333 billion. Eletrobrás and BNDES, with the assistance of foreign consultants and the World Bank, has developed 'rationalization' programmes for distribution utilities, which spells significant labour reductions and privatization of various power generation plants. The oil market has also been de-monopolized, opening spaces for private sector participation and joint ventures with Petrobrás. For natural gas, the Brazilian Congress agreed to make the state monopoly more flexible, allowing local states to grant service concessions to private companies. At present, eleven gas distributors operate in the various states and four others will soon sign contracts to supply to Petrobrás.

Energy policy in recent decades, under the influence of strong sectoral pressures, has led to the transfer of profits between regions, economic sectors and population sectors in the form of tariffs and subsidies. In this way, capital has been drained from the energy sector. Fiscal deficit in turn increased as the national treasury often made up the costs not covered by tariffs, or provided loans for programmes such as PROALCOOL that were not later repaid. It may accurately be said that a process of 'privatization of the state' is under way in Brazil. The political and economic importance of financial groups, such as construction companies, manufacturers of goods and equipment and service companies that provide everything from consultants to third-party mediation, is the most striking aspect of the entanglement of interests, which often entails swindling, over-invoicing and corruption. Big consumers, in turn, are being subsidized, as evidenced by the declining participation of big consumers in power enterprise billing: between 1992 and 1995, billing for high-voltage connections decreased from 44.3 to 40.3 per cent and export incentives were granted to some productive sectors in the form of tariff privileges. Hydrocarbon-producing regions depend on the profits generated by oil and gas and their respective municipalities rely upon the tax they collect from oil operations.

Oil- and natural gas-producing areas are increasingly affected by the risks and environmental disruptions inherent in production and transport technologies. Considerable tracts of land have been lost in coastal and other territories, where fires or crude oil spillage have damaged tourist and historic sites. North of São Paulo and in the north-eastern coastal belt, and in every cane plantation, distillery and power plant run by PROALCOOL, people are living for half of each year with air pollution caused by burning cane-fields and alcohol emissions. The sugar-alcohol industry employs many women and children in harsh, unhealthy jobs with long working hours. Most of these labourers are hired by *gatos* (deceivers) by means of subterfuge, which makes it difficult to enforce labour legislation and improve salaries. In coal-producing areas, mainly in underground mines in Santa Catarina, occupational health hazards and risks are commonplace. In Bage, in the south of Brazil, for example, evidence of air pollution and atmospheric acidity has given rise to claims also on the Uruguayan side of the border. In the southern state of Santa Catarina, the legacy of coal production can be measured in terms of tens of thousands of hectares of lost, barren and perforated land, acidified rivers and underground water contaminated with high levels of heavy metal, and a persistent smell of hydrocarbons, coal tar and sulphurous gas in the air. In places where wood is cut or collected to use for fuel or charcoal production, economic expansion is characterized by environmental degradation. In some cases, trees are harvested for high-quality timber or to clear land for plantations, and charcoal production is a secondary activity carried out by workers illegally hired by contractors and *gatos*. There are numerous cases of entire families working in forced labour conditions, living huddled in forest belts and fenced land, at the mercy of the middlemen of large landowners and charcoal-buyers. In other cases, such as the large eucalyptus plantations, this problem has been only partly solved by the establishment of agro-villages or small towns where everybody works for a single big company (like Carbonita, in Minas Gerais). In these cases, workers coexist with smoke, heat and combustion gases, and are exposed to the risks of falling trees and fire. Hydroelectric dams have flooded over 30,000 square kilometres of land and forests and expelled or 'forcibly displaced' around 200,000 riverside families, depriving them of material and cultural livelihoods. Compensations for displacement were negligible or non-existent, and the resettlement process, if there was one, did not ensure the preservation of former living conditions. Health problems, such as an increase of endemic ailments, have arisen in dam areas, and water quality is worsening with subsequent damage to fishing and agricultural activities and an increased risk of downstream flooding. Large areas of arable soil have been flooded, causing irreversible damage to biodiversity in several cases. Problems generated by hydroelectric expansion have led to the creation of several social movements, such as the National Movement of Victims of Dams. Social inequity is another serious problem in the sector, with more than four million people living without access to electric power. Unemployment is another, since the coal and power sectors and Petrobrás have replaced staff with contractors who generally work under worse conditions for lower pay.

Notes

1. This analysis was prepared by Célio Bermann, Centro de Estudos em Energia e Meio Ambiente, University of São Paulo, Av. Prof. Almeida Prado, 925, Brazil.

2. Bermann, C. (1997), *Projeto Energía e Bancos Multilaterais de Desenvolvimento*, Brazil: Informe Nacional.

15

Colombia[1]

Institutions in Reform

Changes are under way within the institutional framework of the Colombian energy sector, with the state role being limited to planning, regulation and control while private and public companies assume the entrepreneurial lead. Specifically, it is the Ministry of Mining and Energy that is responsible for energy policy, planning, supervision, financing and control of the energy sector via several specialized agencies. An inter-ministerial Regulating Commission of Energy and Gas (CREG) regulates energy markets, controls monopolies, determines transport tolls and tariffs and sets criteria for business management. The Superintendency of Public Services is in charge of consumer protection, but it is proving difficult to protect consumers and defend their rights in a free market economy where enterprises exert pressure to establish and maintain their advantage, so defining regulations for each energy sector is very complex. The present policy allows public enterprises more autonomy, giving them the character of private companies. Meanwhile, new areas are being opened to private investment through privatization and the granting of concessions.

Energy production in Colombia grew 68 per cent between 1980 and 1995, or 17 per cent per capita. Oil derivatives are the most important energy source in Colombia, providing 50 per cent of domestic consumption. Oil and mineral coal are both exported as Colombia's primary and growing energy products. Colombia exports 51 per cent of its total oil production and 82 per cent of its mineral coal, which accounts for 10 per cent of coal sales worldwide. Because of its low price and government promotion, dense oil, which is high in sulphur and nitrogen, is used in industrial boilers. Despite the large availability of oil, there is insufficient refining capacity to ensure a domestic gasoline supply, so Colombia imports this fuel for transportation. Colombia exports its energy resources and imports their derivatives, which constitute its real energy sources. The country's reserves of mineral coal are far greater than domestic demand. Natural gas accounts for 6 per cent of energy consumption and is used mainly for electric power generation and in the industrial sector. Fuelwood accounts for 13 per cent of national energy consumption and is employed almost totally in rural households for cooking. In rural areas, fuelwood is available at no financial cost. Another significant source in the Colombian energy matrix is sugarcane waste, which is used for fuel in sugar-mills and the agricultural sector.

In the 1990s, a slowdown in demand for energy coincided with a shift in the MDBs' policies from power infrastructure construction to sectoral adjustment programmes and electricity market development. New laws in 1994 on public services and electricity ushered in free competition and ended state involvement in the management of energy companies. A policy to foster private sector investments was explicitly formulated in the National Constitution whereby all public or private agents may freely participate in energy operations, except those excluded by law, such as power transmission in the interconnected system. In the electric power sector, the regulatory framework separates generation and distribution functions, which led to the splitting of the state-owned power utility Interconexion Electrica Nacional (ISA) into two independent entities: ISA (transmission) and ISAGEN (generation). The electricity market comprises two categories: regulated and wholesale. The regulated market deals with small and medium-size consumers. The wholesale market covers transactions among enterprises and between these and large consumers, under free supply and demand conditions. It offers two forms: on-spot transactions in the Power Stock Exchange and long-term contracts. The Power Stock Exchange, under the management of ISA, deals hourly with electricity transactions whose quantities and prices are determined by free supply and demand according to predetermined trade rules. Power is transported through the transmission grids of the National Interconnected System. In the oil sector, a state-owned company negotiates several forms of participation in production, leaving exploration risk to private investors. Strictly private participation is not excluded from some phases of the industry, however, and the prevailing feature is a scheme of compulsory association with Ecopetrol. The new mechanisms for oil exploration and extraction are:

Association contracts Exploration companies must carry out all activities at their own risk under an agreed minimal investment plan lasting three or six years. For extraction, the first stage of the contract lasts six years. After field feasibility is demonstrated, there follows a 22-year exploitation period. Five years before its maturity date the contract can be extended for another period of 28 years in the case of liquid hydrocarbons and 40 years for gases. If Ecopetrol does not recognize the field, the associate can exploit it at its own risk and cost.

Contracts with R factor[2] The royalty on productive fields is 20 per cent of the extracted volume valued at the outlet. The remaining 80 per cent (in the case of Association Agreements) is allotted in equal shares to the members. Costs are equally shared. For more than 60 million barrels, after deducting royalties, distribution will depend on the accumulated income and outlay account of the company.

Contracts for gas finding These contracts include a 30-year period for exploitation, with a maximum of 40 years, a six-year exploration period and a four-year retention period. The state enterprise's costs for exploration are completely reimbursed. The associates share the rights and the participation in the operation equally. Development investments are covered by, and belong to, Ecopetrol and the associate company in equal shares, and operational expenses are paid for by both, in the same proportion that is set for distribution of production. The regulatory framework promotes:

- competition (in order to lower market prices);
- more guarantees for the exploration of reliable fields; and
- a change in extracting conditions with larger profit margins for the investors.

Problems associated with this kind of contract include the following:

- the non-renewable nature of resources as well as the environmental impact are disregarded;
- resource exploitation is given a monetarist approach;
- Colombia is forced to export its raw materials; and
- the associate company accelerates extraction.

In addition, experience in the oil sector shows that benefits like employment growth are negligible, royalty sharing is marginal, and the end of concessions leaves behind residual fields that are difficult and expensive to operate.

The entry of new private agents is encouraged, either through the sale of state assets or the opening of new areas for private investment (exploitation concessions, etc.). Public utilities will continue to play a major role, but will tend to operate as private law actors in competitive markets.

Even though the National Development Plan promotes the expansion of the energy sub-sectors to ensure domestic supply, special emphasis is placed on the export of oil and coal. The goals are to intensify and optimize the contribution of energy exports to economic growth and to generate financial surplus for new investment projects. To this end, efforts are concentrated on two fronts:

- increasing oil reserves by fostering exploration, with the limited participation of Ecopetrol; and

- promoting coal investments, strengthening Ecocarbon and promoting special exploration/exploitation agreements with private agents; meanwhile, the state plays an active role in opening and maintaining foreign markets.

The National Development Plan also seeks the construction of a central gas pipeline network to increase natural gas availability, electric power generation and transmission expansion. It likewise seeks to provide energy to rural areas by extending energy supply using appropriate energy resources, which entails the participation of local communities. Energy supply is not equivalent to electrification and, in fact, the proposal is to conserve and substitute commercial fuelwood. A multi-purpose reforestation programme and a plan to improve rural electrification management and coverage complement this.

Traditionally, tariff policies in the Colombian energy sector were designed along political and social lines, with economic considerations being of lesser importance. Energy consumption was subsidized, which led to the inefficient use of energy. In order to rationalize demand and reflect real production costs, a tariff restructuring was planned. This plan was later curbed and applied gradually in view of its social costs and contribution to inflation, and for fear of violent popular uprisings. Some regions have reacted harshly to tariff increases and in the last few years there has been a great increase in the number of enterprises defaulting on energy bills, particularly in the electricity sub-sector. Because the Colombian government did not dismantle energy subsidies rapidly enough for the World Bank, the last disbursement of its adjustment loan for the electric power sector, US$75 million out of US$300 million total, was cancelled. In the major cities, the poorest 10 per cent of the population spends up to 11 per cent of its income on energy, while the global population averages 2 to 3 per cent.[3] The highest energy consumption is for cooking. The unequal impact of energy costs on home budgets is manifest in the different degrees of access to electricity: in the richest 20 per cent of the population, 98 per cent of families have electric power service; in the poorest 20 per cent, 18.7 per cent of families do not have grid access. The cost of access to electricity and natural gas differs according to the social and economic characteristics of the several regions of the country. The new policy aims at organizing a cross-subsidy system so that the highest-income segments of the population pay a share of the subsidy to the poorest segments. In previous years, the government has exclusively provided this subsidy.

Colombian energy policy makes no provision for sustainable use of energy resources through conservation or transformation as the basis for national development. Colombia uses energy inefficiently and domestic demand exerts a high pressure on renewable and non-renewable natural resources alike. Energy conservation and energy source substitution by less costly fuels are goals of the National Development Plan, which focuses on two fronts: demand-side knowledge improvement and demand-side management adjustment. Two kinds of tools are advanced for the latter: appropriate price and tariff signals, and specific programmes for efficient fuel use supported by research and development activities. There are also programmes for substitution of some low-efficiency fuels such as fuelwood. Programmes for reducing its use have been implemented without much success since they were not well suited to people's needs. Electricity, especially in homes, is used inefficiently. For cooking and water-heating, more efficient fuels could replace electricity. That is precisely the idea of the Plan for Mass Use of Natural Gas, but there are two basic problems with its implementation: cultural patterns and the cost of new stoves and installations required for the shift from electricity to gas. Culturally, the use of electricity for cooking and water-heating is a status indicator in Colombia. Also, installation of new gas networks, inside and outside buildings, is done without aesthetic considerations, which are important especially in high socio-economic sectors. There are also important losses, and thus low efficiency, in the consumption of sugarcane waste. In the case of oil, discussions have focused mainly on how to attract foreign investors by reducing oil taxes and royalties, which are currently higher than anywhere else worldwide. This is mainly because the so-called oil rent is a major source of state revenue upon which the country is heavily dependent.[4] The Ministry of Mining and Energy meanwhile regards environmental licensing by the Ministry of Environment as a barrier (a bureaucratic hindrance) and a restriction rather than as a necessary planning tool.

Lack of Effort for the Support of Sustainable Energy

The environmental and social impacts of the Colombia energy sector projects occur at the exploration, construction, exploitation and transportation stages where impacts are concentrated in rural areas, and at the consumption phase where impacts are concentrated in urban areas. The main environmental impacts of natural gas and oil exploitation, production and transportation are:

- soil degradation, manifest in erosion processes, arising from the construction of access railroads and camp-sites, and from soil-levelling for well-drilling, hydrocarbon production and transportation;
- alteration of water quality as a result of liquid discharge and underground or surface water pollution caused by oil spills and dangerous waste pollution;
- atmospheric pollution produced by gas-burning in some fields;

- increasing deforestation and depredation of unique ecosystems, many of which are strategic;
- loss of biodiversity;
- disruption of delicate food chains with slow recovery capacity (mainly in the Amazon basin); and
- contamination through tank infiltration as well as through oil and solvent discharge.

Oil refining is identified as one of the most environmentally harmful activities because it pollutes the air and generates large amounts of chemical wastes with varying toxicity levels. In Colombia, oil is still used in boilers because of its low cost and the government promotes 'Castilla crude', a very dense oil with high levels of aromatics and unusually high percentages of sulphur and nitrogen, for this purpose. Since this oil was very difficult to sell abroad, it was promoted in the domestic industrial market as fuel, even though it has severe environmental consequences due to emissions of sulphur and nitrogen oxides. In relation to the country's coal extraction transportation the main environmental impacts are:

- in the air: high levels of coal dust both inside and outside the mines;
- in the water: pollution of rivers near to the exploitation site by solid particles and soluble minerals; and
- in the ecosystem: deforestation of the forest cover (6,000 ha for medium-size and small mines) with subsequent loss of wildlife.

Underground mining, because of low technology, the lack of resources and the use of unskilled labour, causes heavy impacts, including:

- destruction of vegetation and soil damage;
- activation of erosive processes;
- destruction of wildlife;
- pollution of underground water, which filters into the mines and is pumped into the rivers without treatment;
- increased methane concentrations from explosions due to methane–oxygen mixtures;
- air pollution by particles in suspension that cause or contribute to respiratory ailments;
- increasing sediment load in watercourses; and
- high level of accidents.

In hydroelectric generation, impacts on the physical environment are basically related to changes in flow rates of surface and underground watercourses, and, in the case of dams, also include soil destabilization. The occurrence of these impacts is highly probable; in the majority of cases, the opportunity for mitigation is small or non-existent; and the impacts generally affect a wide range of activities. Some of the impacts are:

- soil erosion;
- gradual loss of capacity of dams affected by erosion and sedimentation, which in turn leads to a decrease in the dam's useful life and in generation capacity;
- downstream erosion;
- in some cases, a decrease in water flow in shallow parts of rivers and increased water flow in recipient channels; and
- production of hydro-sulphuric acid in several dams, because of non-removal of the existing vegetation prior to the flooding of the area and the quality of contributing water.

Impacts on the biotic environment are related to the alteration of forest and aquatic ecosystems as a consequence of the establishment of dams. The main impacts are:

- destruction of wildlife due to flooding, break-up of established biological relationships and the loss of biodiversity;
- removal of covers of primary and secondary forests, stubble-fields and grass;
- destruction of habitats of biotic communities;
- discontinuation of fish migrations; and
- substitution of aquatic habitats.

The major impacts of thermoelectric generation are:

- related to fuel storage and transportation: particulate emissions and contamination from spills;
- related to burning: emission of oxides of sulphur, carbon and nitrogen oxides, unburned hydrocarbons, ash and slag production; and
- water pollution: hot water waste, or thermal pollution, emissions of oily wastes.

In fuel use, particularly in urban areas, complete and incomplete combustion of fossil fuels for transport and industrial purposes contributes to air pollution derived from:

- gasoline and other derivatives;
- fuelwood;
- cane wastes;
- coal; and
- LG and natural gas.

High emissions of carbon monoxide (CO), sulphur dioxide (SO_2), volatile hydrocarbons and nitrous oxides (NOX) from cars are due to:

- the types of gasoline used in Colombia in terms of sulphur, olefins and oxygenated compounds, among others;
- the age of the vehicle stock;
- deficient car maintenance; and
- inadequate burning technologies.

The Uwa, a native community, have tried to prevent oil exploitation in their habitat for cultural reasons, and threatened to commit mass suicide if exploitation begins. The socio-economic impacts stemming from hydro-electric generation, especially where dams are associated, are severe and affect large areas. It is usually difficult to prevent or even mitigate these impacts, which include conflicts arising over traditional land use and displacement of populations due to flooding.

Alternative forms of energy that are less damaging to the environment have received little promotion other than the National Development Plan, which has established the use of alternative energy sources as a priority whenever these are economically feasible. But in practice, little effort has gone into making these options more economic. Institutional support for these sources is concentrated in the INEA, an entity that does not have an adequate staff and budget to undertake this task – goodwill alone has not transformed intention into reality. Studies on alternative sources are highly inadequate. The Feasibility Study of Project Azufral, for example, examines the potential for geothermal energy, and is financed by the Inter-American Development Bank (IDB). It examines the country's geothermal potential in the area of Nariño, with a view to improving the reliability of the country's generation system, reducing the region's energy deficit, improving the planned energy integration between Ecuador and Colombia, and encouraging private sector participation in the development of this natural resource. To improve energy efficiency and foster non-conventional sources in Colombia, technological research of alternative energies and implementation of energy systems based on indigenous resources are needed. A Rational Use of Energy (URE) programme is being implemented within the framework of an IDB-financed project, and has the following strategies:

1. Demand-side orientation: this comprises a set of actions to induce consumers to improve their consumption habits and employ energy-efficient equipment and processes. This strategy is developed through the following activities:

- design and implementation of incentive schemes to promote URE: tariff options with a clear signal of energy use will be considered, as will

schemes for improving the efficiency of cooking and electrical appliances;
- identification of substitute processes and evaluation of savings potentials in consumption, to obtain global aggregate numbers for savings potential;
- training and human resource programmes of the URE, through a national or institutional formative plan; and
- educational campaigns for the efficient use of energy.

2. Optimization of the generation processes.
3. Fuel substitution, including partial substitution of natural gas (NG) or liquid gas (LG) for electricity in residential and commercial sectors; alternative fuels (LG or pressurized NG) for gasoline and diesel for public transport; and LG for fuelwood in rural areas. Some specific proposals include the use of heavy oil as asphalt adhesive for paving as an option for use of Castilla crude oil and the commercialization and use of coal briquettes in pilot projects to offset fuelwood consumption in the rural residential sector. Note that several of these proposals regard fuelwood as an inappropriate fuel without acknowledging that it is renewable, unlike LG or coal, the target substitutes. Up to a certain point it can be said that the short- middle- or long-term environmental effects are not being considered.

The Banks' Shift to the Private Sector

Over the last 50 years, the MDBs have developed several projects in Colombia, nearly 40 per cent of which are in the energy sector, with a concentration of loans in the electric power sub-sector. Hence this sub-sector is responsible for a significant proportion of foreign debt. This source of indebtedness is shifting from projects to state and sectoral reforms. It is no coincidence that this shift from project to sectoral loans is fostered by the MDBs. Nor is it a coincidence that actions taken by the Colombian government are in line with World Bank and Inter-American Development Bank policies, including modernization, privatization of services, and establishment of state agencies for regulation and planning. According to the Bank itself, 'Great consideration was given in several projects to the strengthening of regulatory frameworks for specific sectors, mainly for those recently privatized. Such is the case, for instance of the project for the development of an electricity market in Colombia.'[5] There are no projects related to the development and use of alternative energies. Only the Rational Use of Energy Project includes a feasibility study for a geothermal energy plant, which, being a reimbursable loan from a Japanese institution, will be developed by Japanese consultants. There are no larger studies related to alternative energy sources. The number of loans granted to state agencies has decreased in favour of loans to the private sector. In the case of the World Bank, these are granted through the International Finance Corporation (IFC), or directly by the IDB. The IFC, for example, is developing Project 7283 with the Colombian company Promigas, which involves a three-year expansion plan for this company. The amount of the loan is unknown. The project qualifies as environmental Category B (those for which the WB requires an Environmental Analysis) and a course of action was suggested in accordance with the potential impacts foreseen in the studies. It is worrying that opposition to these projects may become impossible as a result of their being privately undertaken. For the time being, difficulty in obtaining relevant information about these loans hinders all research on the subject. The IDB is financing the Rational Use of Energy (URE) project. The loan amounts to US$11 million, of which US$10 million is earmarked for programme development, and US$1 million will be provided by the Japanese Fund for Consultant Agencies as a non-reimbursable credit. This latter amount is for the feasibility studies of Phase I of the Azufral Geothermal Field Project. The total cost of the URE project is US$13.3 million, of which US$2.3 million will be provided locally. The project does not include significant social considerations. Only in a few cases is reference made to indirect benefits to society. The only environmental consideration seems to be focused on reducing the consumption of heavy fuel and its partial replacement with natural gas, which would – according to the Bank – decrease the overall emissions of nitrogen and sulphur oxides and would also reduce carbon dioxide emissions and the associated global warming effect. Positive as this might be, it is clearly insufficient regarding the complexities of the energy problem in Latin America, particularly in Colombia.

Notes

1. This chapter was prepared by Hildebrando Vélez, Tatiana Roa, Marta Rincón and Diana Milena Guio, CENSAT 'Agua Viva', Carrera 19, No. 29 – 120.202, Apartado AÈRO NO. 16789, TELEFAX: 57-ñ2456860, Santafé de Bogotá, Colombia.

2. The 'factor R' agreement states that the production of hydrocarbons (coal, oil and/or gas) is divided; the factor is calculated by dividing the partner's income (as a result of production) by the invertment costs, after deducting the repayments which are made by Ecopetrol. The factor comes into play when the accumulated production of a field reaches 60 million barrels.

3. May, Ernesto (general coordinator) (1996) *La pobreza en Colombia. Un estudio del Banco Mundial*, Tercer Mundo Editors, January.

4. The total amount of all the economic resources that remain in the country as a result of hydrocarbon exploitation, such as royalties, direct or indirect taxes, remittances and transference payments.

5. World Bank (1996) *Annual Report*, Washington, DC: World Bank.

16

Mexico[1]

Energy Sources

Traditionally, the institutional structure of the Mexican energy sector has been dominated by two large state-owned companies: the oil company Pemex and the Comisión Federal de Electricidad, or CFE, which along with Luz y Fuerza del Centro, another public electric utility, control the operation of the national power systems and electricity services. The semi-official Secretariat of Energy, Mining and Industry (SEMIP) is responsible for national energy policy, while the Treasury and Public Credit Secretariat (SHCP) allots financial resources, set tariffs and imposes taxes on energy. An inter-ministerial commission (Group of Fuel Policy) undertakes strategic planning for the energy sector, coordinates investments and evaluates environmental impacts. The Energy Regulatory Commission (CRE) was established in 1995 as an autonomous government body. A trusteeship (FIDE) and an inter-governmental commission (CONAE) are oriented towards energy conservation and efficiency. Pemex, or Petroleos Mexicanos, monopolizes all phases of oil and gas production and commercialization and its vertically integrated operations include Petroquímica Basica (Basic Petrochemicals).

Until recently, only the state was empowered to generate, distribute and store electricity, oil and its derivatives. Recent amendments connected with the integration of Mexico into the North American Free Trade Agreement (NAFTA) enabled private sector participation, especially in power generation and gas transportation and storage.

From 1980 to 1995, the total energy supply grew by 115 per cent, or 59 per cent per capita. Mexico has a substantial energy resource base, with large oil, natural gas, thermal coal and water reserves as well as a rich solar and wind energy potential. Since the 1970s, however, when some of the largest oil and gas reserves in the world were discovered in Mexico, primarily around the Gulf of Mexico, it has relied mainly on hydrocarbons for 82.9 per cent of gross domestic supply of primary energy. The discovery of oil changed Mexico's status from a net importer of hydrocarbons to a net producer ranked as seventh world hydrocarbons producer and tenth crude oil exporter. Pressed by its foreign debt, Mexico has tried to maximize oil exports to improve its payment balance. Gas production and consumption rates a distant but significant second to Mexico's oil industry, but it is still among the world's largest producers and consumers of natural gas.

Only 34 per cent of the country's hydroelectric potential is being met, and its contribution to overall consumption is low, though present generation capacity would double if the identified projects were implemented. Most hydroelectricity is generated in the least populated south-east of the country, and a large proportion is transported to the central and northern regions. In the last decade, the importance of oil-fuelled power plants has increased as the contribution of hydroelectric plants has declined.

Fuelwood is mainly used in rural communities for cooking, and sugar-cane waste is used in sugar-mills. Biomass accounts for 5.9 per cent of the gross domestic supply of primary energy. Mexico has uranium reserves and several nuclear plants.

New Regulatory Frameworks

The Mexican constitution defines the parameters for state and private capital involvement, with section 27 stating that land and water belong to the nation, which possesses the right to transfer ownership to private entities. It also establishes the right of the state to impose conditions and regulate private property in the best public interest. It provides that the exploitation of natural elements can be taken over, with a view to a fair distribution of public wealth. Until 1995, hydrocarbon production, distribution and storage, as well as electric power generation and distribution, were state-run activities. After Mexico's adhesion to NAFTA, the barriers for private participation in electric power generation were removed, but electric power commercialization still remains under the CFE, a vertically integrated public utility in charge of power transmission and distribution. Therefore, possibilities have been created for private sector involvement in power co-generation, either through subsistence production or independent production, but there is not a wholesale electricity market. Likewise, natural gas transportation and storage have been opened to private-sector participation.

The provisions of the Energy Regulatory Commission (CRE) on electricity and natural gas markets are focused on the relations between private agents and the state, and especially on supply agreements between independent producers and CFE. But there are important weak points in the regulation of electric power procurement by independent and small producers. The regulation trend is that higher electric power tariffs encourage private sector participation in generation. The new provisions on electricity exportation/importation – which are less restrictive than those established by NAFTA since they do not require the intervention of CFE as NAFTA does – could lead US companies to extend their power systems through installations inside the Mexican border that would not be connected to the Mexican national grid. In fact, southern US electricity companies may be interested in installing power generation plants on Mexican territory in order to profit from the economic advantages granted by Mexican environmental provisions, which are less strict than those of the USA, and export electricity through their own transmission lines in Mexico. In the oil sector, where barriers to the entry of new agents persist, the new regulations allow Pemex to develop a policy of international associations. This company's bias towards forming strategic alliances is revealed in several initiatives:

- a strategic association with Shell to increase Pemex refining capacity, purchasing 50 per cent of a refinery in Texas;
- establishment of Mexpetrol in association with private capital, to export services and goods in oil operation projects and in the petrochemical industry; and
- purchase of 3.5 per cent of the stock of Spanish Repsol and the creation of an association for projects, co-investment and oil supply agreements, which has opened new market opportunities for Pemex in the European Union.

Mexican energy policy still prioritizes oil resources and oil income as the backbone of fiscal income. Extension of the oil platform is a high priority, while Pemex's activities in hydrocarbon refining, processing and marketing, even first-hand sale, are encouraged. At the same time, policy aims to reduce oil consumption, which is currently at a 2:1 ratio compared to natural gas power generation, reversing the present relation by increasing natural gas consumption and establishing environmental norms in critical urban areas and in the rest of the country. The strategy consists of the following steps: investing in substitutes for oil, prioritizing the use of natural gas and seeking alternative outlets for oil displaced by the introduction of natural gas.

Although Mexico has natural gas reserves, it must develop the fields as well as the transportation and distribution infrastructure to supply gas to consumers. At the same time, the conversion of two-thirds of CFE thermoelectric units from oil to natural gas has been planned. The growing interest in gas can be partly explained by the aggressive pressure exerted by the Canadian enterprises Nova Corp and TransCanada, which, for several years, have tried to form joint ventures for gas production and distribution as well as for gas-fuelled power generation. As for electricity generation, there are plans to develop hydroelectric projects. No clear policy has been devised for the development of other energy sources.

Enterprise and Market Restructuring

The integration of Mexico into NAFTA has broadened the field for private sector involvement in electric power generation and gas transportation and storage through constitutional and legal amendments. Privatization has been accompanied by cuts in employment, salaries and benefits, mainly in the case of oil-workers. The privatizing proposal has met with resistance in government, where some groups are not convinced of its benefits because they regard energy as a strategic sector that should remain under state ownership. Mexican society is generally opposed to privatization after 15 years of privatization that produced wealth and benefits mainly for the friends and comrades of the president. At the same time, there are also powerful interest groups around the state monopoly of CFE who control resources and hold considerable power. Given the nature and range of energy sector operations, many opportunities are open for corruption and fraud and this has deterred investors. This has not been the case with more clear-cut schemes, such as the sale of the petrochemical company and of CFE, however. So private investment has focused on the distribution of natural gas and foreign investors do not appear eager to invest in the Mexican power sub-sector.

The World Bank-promoted Energy Regulatory Commission (CRE), formed in 1995 to assist private sector inclusion, is nominally in charge of tariff setting, but in the case of electricity, the Treasury Department sets tariffs for CFE. This arrangement has created conflict since the Treasury bases its tariff-setting policy on macro-economic and fiscal balance, or even political considerations that do not necessarily reflect the company's situation or the needs of consumers. Electricity and oil tariffs in Mexico are well below real costs. The federal government, through state-owned companies, has subsidized electricity and liquefied gas for decades on the grounds that they are basic human needs. But subsidies could not be sustainable for an unlimited period of time, and encouraged the non-rational use of energy resources. At the end of the 1980s, the government, together with the World Bank, began to restructure the tariff system to remove consumer subsidies. Recurrent crises have hampered its success, however, and although a steady tariff increase has taken place, CFE tariffs still do not cover costs. Every time tariffs approach real cost levels, a new devaluation occurs that widens the gap, and subsidies rise.

Subsidies especially benefit big consumers and the richest areas of the country. The tariff system is quite complex, being divided by sector and by level of use on the grounds – according to CFE – that this protects consumers in areas without local resources for generating the power they consume. This policy has caused political strife: the peasants of Chiapas, one of the poorest Mexican states, for example, complain that they are subsidizing consumption in the rich northern states. The new tariff policy has been developed in view of private participation in the sector. To this end, tariffs will be raised to cover costs, and subsidies will hence be reduced.

Lack of Finance for Sustainable Energy

Energy conservation and efficiency are considered by some programmes promoted by CFE and by agencies oriented towards this end, such as FIDE and CONAE. Budget availability, however, is constrained, compared with the proposed objectives. The economic crises and lack of political will also limit the allotment of resources to government and NGO programmes, that promote the use of renewable energy sources and energy efficiency improvements. CFE has hindered the development of renewable energy by allocating available resources to new conventional plants without considering the kind of technology used. Alternatives such as solar and wind power are often unprofitable in the short term and CFE fears renewable energies as they may reveal its own inefficiency and that of its thermoelectric units.

The environmental impacts most studied are those generated by hydro-carbons production, transportation and industrialization, and those arising from oil derivatives, such as gasoline and oil fuel. Negative impact on the shallow section of the Coatzacoalcos river is internationally regarded as one of the biggest environmental problems generated by the oil and petro-chemical industries. Fishing, agricultural and cattle-raising activities have become impossible in the area, where the first refinery was established in 1907 and 65 giant petrochemical stations have been established since, making this one of the most important petrochemical centres of America. In the nearby state of Tabasco, crude oil extraction has already damaged more than 90,000 ha of land and water. Problems include heavy metal contamination, acid rain, and water and land salinization, leading to health problems and serious degrading of the regional ecology and economy. Lack of maintenance of installations has led to a rise in the number of accidents in the last five years corresponding with the massive retrench-ments of skilled workers who warn that over 8,000 km of pipeline and 110 installations are virtual 'time-bombs'. But the pressure for accelerated oil extraction to meet growing demand in Mexico and the United States is at the root of the problem and jeopardizes the stability of reserves.

Mexico has significant hydraulic resources and hydroelectric energy has a great potential, but the way in which CFE has built and operated its hydroelectric plants has had a negative impact on local people and the environment. The construction of several dams has led to the displacement of people and severe social problems in the Laguna Verde area. This experience gave rise to peasant and native movements that resisted projects, some of them financed by the World Bank. For three years, peasants in the south-eastern states have opposed dam management that threatens to flood their subsistence agricultural lands.

The impacts of electric power generation include those of the nuclear power plant at Laguna Verde and several hydroelectric units. Nuclear reactors have also had a negative impact on the environment through radioactive water discharges into small lakes and lagoons, leakage of radio-active gases and liquids and the discharge of tens of tons of radioactive wastes annually, including lethal plutonium.

Power tariffs are an important factor in terms of their social impact. In the state of Chiapas, CFE has cut off the electric power supply to hundreds of communities (and thousands of users). These users, either small enter-prises or people, some of whom are among the poorest in the country, applied for special tariffs because their water resources constitute an import-ant source of hydroelectric energy for the country. Furthermore, for the last ten years CFE has used land in Chiapas to build dams but has failed to pay land taxes. According to CFE reports, 131,000 users from the state of Chiapas have become indebted to the company. The vicious circle of default on payment, power cuts and protests has reached crisis levels. In response CFE launched 'Una Luz Amiga' (A Friendly Light), a programme that provides a discount to 82 per cent of households in the state. Those who consume up to 10 kW/hour have a 50 per cent discount on their bills. Additionally, the government of Chiapas has spent more than 44 million pesos (approximately US$5.8 million) to cover part of the electric energy bill of the users (*La Jornada*, 25 October 1996). The issue has raised angry protests from the population of Chiapas and forced the authorities to modify some of their operational policies. These measures have not solved the problems, however, so massive demonstrations (65,000 farmers in 1996) have continued and several resistance organizations have been formed by local farmers and supported by researchers, non-governmental organiza-tions and opposition parties.

The high relative cost of renewable energy still hinders its expansion in Mexico. With the exception of large-scale hydroelectric projects, renewable sources are mainly used in small-scale pilot projects. As long as oil is available, the official tendency is to use it on a short-term basis. With longer-term analysis, however, some renewable sources are becoming competitive with conventional ones. Some outstanding projects based on renewable energies exist and have attracted the corporate interest of, for example, US Amoco in association with Enron, Soldar Power Development, Bechtel and others interested in promoting renewable energy in communities of between 70,000 and 140,000 that are not on the electricity grid.

Solar energy Mexico occupies a region of the world where the sun is at its greatest intensity. For that reason, it has been the site of several solar thermal and photovoltaic studies. In 1991–94, more than a hundred rural communities were electrified by means of photovoltaic cells, with a power capacity of around 1,500 kW. It is thought that over 50 per cent of the rural communities that are not yet connected to the grid could benefit from these energy sources.[2] Mexico has installed an estimated 40,000 solar panel systems used for electricity generation (photovoltaic systems), water-heating and water-pumps. An example is the project Electricidad Rural con Energía Solar (Rural Electricity with Solar Energy). This project was originated in CFE, and is conducted by the Secretariat for Social Development (SEDESOL). It promotes the use of solar energy by means of solar roof collectors for lighting in 10,000 communities in several regions of the country.

At the same time, SEDESOL, with World Bank assistance, is implementing a project to support poor districts. This project – a sort of Social Investment Fund (FIS) – offers a menu of public works that may be carried out with the assistance of SEDESOL and the World Bank. Electricity is one of the prospective services, but it can only be supplied from the electric power grid. Most of the funds for this project are assigned to rural districts, but rural communities are not allowed to choose projects that involve non-traditional electric power supply, even though these may be more efficient and cost less.

Wind energy Windmills for water-pumping are widely used in rural areas, particularly in coastal regions. In recent years the national energy balance reports 330 kW power capacity in air-generators and water pumps. The estimated potential for wind power in Mexico is 2,000 MW, 50 per cent of which is concentrated in the Tehuantepec isthmus.[3]

A CFE pilot wind project in Ventosa, Oaxaca, supplies 1.6 MW of electric power. Recent studies reveal that this region has a wind potential of up to 600 MW. Recently, the Danish company Vestas proposed to CFE a wind project of 300 MW in Ventosa and offered to guarantee the sale price of electricity at US$0.04 per kW/h – with the average price for the sector being nearly US$0.06 per kWh, this is a cost lower than the average

0.06–0.08/kWh for wind projects worldwide.[4] Unfortunately, the CFE said it would accept the project only at 0.02/kWh: an unrealistic price for the Mexican market.

Solid waste Solid waste may be used to generate electricity from biogas produced in sanitary back-filling or directly from its burning. The estimate of electricity potential of this sort of technology in Mexico is 300 MW, using garbage from the country's largest cities.[5] The cost of a plant of this type is double that of a conventional power plant.

Biomass Biomass potential has not been quantified in Mexico. Fuelwood consumption in rural communities and waste from the sugar sector produce 4 per cent of total energy. The existing capacity of power-generating facilities using sugarcane waste and oil is nearly 400 MW. Plant factors are estimated at 15–20 per cent, however, since they operate only in the harvest season.[6]

Energy efficiency With currently available technology, Mexico's energy conservation could reach 20 per cent through simple administrative reforms. The government and some NGOs with small alternative projects are carrying out programmes aimed at encouraging energy efficiency and the use of renewable energy sources. Although most of these programmes are relatively small, there is technical and institutional capacity to expand them. The economic crisis, combined with a lack of political will, however, has limited the availability of resources for such purposes. If these programmes are to have a significant impact at the national level, they must be integrated into mainstream energy sector policies and projects. They are still seen as 'pilot projects' or 'alternative projects' rather than as part of a global strategy for the energy sector.

Apart from some efficiency and conservation programmes promoted by CFE, there are two government-sponsored programmes that promote efficient use and rational consumption of energy. One is a private trustee-ship established by CFE, and the other is an inter-ministerial agency. These agencies have a high capacity to promote innovative programmes at the national level, but their budgets are still relatively limited. FIDE – the Fideicomiso de Apoyo al Programa de Ahorro de Energía del Sector Eléctrico, or Supporting Trusteeship to the Energy Conservation Programme of the Electricity Sector – was established in 1990 by the CFE and the Compañía de Luz y Fuerza del Centro (CLyFC), as a private agency whose aim is to encourage the efficient use of electricity. In the last seven years, FIDE has assisted some 50 projects in industrial installations and another 85 in commercial enterprises, and financed 90 public lighting projects.[7] FIDE facilitates cooperation between public and private companies and trade unions. Between 1990 and 1995, FIDE projects resulted in an estimated 5,400 gWh savings (nearly 5 per cent of the national electricity consumption for that period), reducing the need for new generating capacity by nearly 180 MW.[8] FIDE plans to increase this saving to 12 per cent of national energy demand for the year 2000. According to FIDE, its programmes resulted in a reduction of sulphur dioxide, nitrogen oxide and carbon dioxide of over 420 million tons.

The Comisión Nacional para el Ahorro de Energía (National Commission for Energy Conservation – CONAE) is an inter-ministerial agency that coordinates the electricity conservation policies of the federal government. The commission's initiatives include the design of minimal energy consumption standards for producers. These standards are expected to enable an annual saving of 40,000 GWh by the year 2004, avoiding the need for 2,000 MW of new generation capacity. According to CONAE, the establishment of measures to improve consumption efficiency will enable a 15 per cent saving of the energy consumed by industry, the federal government and public lighting. According to the Energy Secretariat, the set of actions to promote energy efficiency will yield a 5,513–7,951 GWh/year saving by 2000.

Bank Failure on Social and Environmental Commitments

In the energy sector, the World Bank seems to have played a more significant role than the Inter-American Development Bank (IDB) in recent decades. Since the 1980s, both banks have curtailed financing for large infrastructure projects in Mexico. Instead, they have favoured sectoral and structural adjustment programmes of the Mexican economy and social programmes for poverty alleviation. Although energy is not a priority for the banks in Mexico, significant resources have been channelled to the sector. The World Bank has provided energy project loans of nearly US$960 million since 1990, while the IDB has financed two projects for a total of US$405 million. It seems that the World Bank sets the priorities for the Mexican energy sector and the IDB follows suit. Both banks have channelled most of their energy projects (96 per cent) to the electricity sub-sector, mostly for sector restructuring (including important investments in thermoelectric plant maintenance) and for power generation by thermoelectric and hydroelectric plants. The World Bank has claimed that these projects are devoted to demand-side management, but in fact they are restricted only to modifications in energy pricing. Both banks say they have no investments in the oil sector, but have undertaken a 'counselling role'. The World Bank has produced non-public documents recommending measures for both the oil and natural gas sectors and has financed studies undertaken by private consultants for restructuring the sector. In 1990, the Bank proposed the following long-term strategies for its interventions in the Mexican energy sector:

1. assisting Mexico to raise foreign currency to finance its investment programme in the sector;
2. promoting finance policies that exclude the need of government subsidies and ensure the efficient use of resources;
3. fostering the opening of the procurement process to international competition;
4. helping CRE's attempts of institutional strengthening;
5. guaranteeing the consideration of the minimum cost principle in the energy sector programme;
6. strengthening norms and proceedings on environmental and social issues in both construction and operation projects; and
7. promoting appropriate policies for the promotion of co-generation and the efficient use of energy.[9]

A review of the Bank's investments in the last decade shows that the first four goals are its real priorities. The achievement of the stated objectives of efficiency and social and environmental issues has been cut short. Loans granted in recent years have been based mainly on supply considerations, plant maintenance and construction to meet the increasing energy demand. At the same time, in political negotiations the banks are pressing the government to restructure the sector by changing the regulatory system and opening the sector to private investment. Under the strong influence of the MDBs, the Mexican government has encouraged gradual privatization of the energy sector. The Electricity Law as amended in 1992 enabled private sector participation in power generation. In December 1994, the World Bank approved a loan earmarked for Technical Assistance for Infrastructure De-incorporation (or privatization). Within the framework of the technical assistance project, the World Bank insisted that CFE be split according to its three functions (generation, transmission and distribution) as a first step in restructuring, which clearly paves the way for privatization. The World Bank also insists on the need to raise electricity tariffs in order to bring them into line with long-run marginal costs.[10]

Policy papers such as *The World Bank's Role in the Electric Power Sector* (1992) seem to be the most important for Mexico. The Bank has put an emphasis in these policies on the establishment of a transparent regulatory framework that would help the sector operate under strict rules and reduce corruption. This would create more opportunities for projects based on renewable resources, efficiency and conservation. The principles contained in *Energy Efficiency and Conservation in the Developing World* (1992) are more innovative, but there is no evidence that they are being implemented in Mexico. The principle that the Bank should include the subject of energy efficiency in policy dialogue with the government does not seem to be adhered to. Dialogue on energy efficiency has been mentioned only briefly in the context of privatizing the sector and liberalizing tariffs.

Notes

1. This chapter was prepared by Mary Purcell and Susana Cruikshank, Equipo Pueblo, Francisco Field Jurado #51, col. Independencia 03630, Mexico.

2. Arriola Valdés, Eduardo (CFE) (1995) 'Planeación del sector eléctrico en México en el nuevo marco regulatorio', in *Integración de Mercados*, Mexico: UNAM.

3. Ibid.

4. Interview with Dr Manuel Martinez, Energy Research Center, March 1997.

5. Arriola Valdés, 1995.

6. Ibid.

7. Silver, Daniel (1995) 'Mexico's FIDE: A leading light for energy efficiency', *E-notes*, vol. V, no. 1, January–March, p.5; and Cardona, Carlos, (1996) 'Mexican energy efficiency: public interest and private initiative', *E-notes*, vol. VI, no. 2, November, p. 4.

8. Biller, Dan and Suzanne Maia (1996) *Pursuit of Sustainable Energy Development in the Americas: A Look at Recent Progress*, World Bank, November, p. 18.

9. *Staff Appraisal Report: Mexico Transmission and Distribution Project*, 20 March 1990, pp. 21–2.

10. Along these lines, part of the transmission and distribution bank loan, approved in 1989, was to help in the adjustment of prices.

17

Uruguay[1]

The Institutions

For much of the twentieth century the Uruguayan energy sector was dominated by two national utilities, of which one prevails in the electricity sector and the other in the oil sector.[2] Not only did each of these utilities hold a monopoly on their respective sub-sector, but they also regulated them. Private sector participation was restricted to the realm of the distribution and retail of liquid derivatives and liquefied gas. The concession for the provision of gas to the Uruguayan capital, Montevideo, belongs to a company that was alternately privately and publicly owned and is now controlled by French capital, GASEBA. A binational (Argentina–Uruguay) hydroelectric project (Salto Grande) was built 20 years ago and a Binational Technical Commission manages the power generator. This institutional structure is, at present, undergoing deep changes. Design and implementation of energy policy is the responsibility of the Ministry of Industry, Energy and Mining (MIEM), which operates through the National Energy Directorate. The National Load Dispatch, an agency under the National Energy Directorate, controls the technical operation of the national power grid. The Planning and Budget Office, also at ministerial level, supervises and plays a decisive role in investment, tariffs and indebtedness of public utilities. The new Electricity Regulatory Framework Law allows private sector involvement in power generation (and possibly also distribution). It has created a wholesale market under the management of the Wholesale Electricity Market Administration, in which both public and private operators will participate. In the oil sector, ANCAP is forming associations with private companies. The expansion of Argentine natural gas is bringing new actors to this institutional scheme: transnational consortiums, some in association with ANCAP, will exploit gas transportation and distribution throughout the country. Ever since 1931, the Uruguayan oil sub-sector has been run by the monopolistic, state-owned ANCAP, which is the legal authority for the refinement of crude oil, and the import of liquid, semi-liquid and gaseous fuels. The distribution and marketing of liquid derivatives are also partially implemented by ANCAP, which sells to a select few large consumers. Most of this business is in the hands of private distributors, as is distribution of liquefied gas. For the time being, ANCAP regulates the distribution and commercialization of oil and gas without MIEM intervention through confidential agreements with private distributors. The gas sub-sector was restricted to the operations of GASEBA, the company that distributes gas in Montevideo under a monopolistic concession.

Energy Sources

Without its own commercial reserves of fossil fuels, Uruguay relies on oil derivatives to meet an energy demand that has grown for the last 15 years. Oil derivatives are the primary energy source, accounting for 57 per cent of end-use consumption. Localized sources, such as hydroelectricity and fuelwood, have been developed intensively to supply 19 per cent and 22 per cent of end use respectively. But now the potential for large-scale hydroelectric generation is almost exhausted, with two large projects, Palmar and Salto Grande, built between 1974 and 1983, and the Rincon del Bonete and Baygorria dams. However, the potential exists for small hydropower projects to be developed in some places. Fuelwood consumption expanded rapidly following the oil crises of the 1970s, from traditional household combustion to widespread industrial use. This trend of substituting fuelwood for hydrocarbons in the industrial sector is being replaced with gas. Argentine natural gas is incorporated into the Uruguayan energy matrix for power generation and residential consumption. The gas pipeline that connects the Argentine province of Entre Rios with the Uruguayan city of Paysandu is already operating, and the construction of the Buenos Aires–Montevideo gas pipeline is on course.

Market Deregulation, Regional Integration

In view of market deregulation, which will likely be required by regional integration, ANCAP is making strategic alliances to operate under competitive conditions. The expansion of the gas sub-sector through the inclusion of Argentine natural gas has brought new actors and new regulations that are not yet enforced. In the electric power sub-sector, a new Regulatory Framework Law to establish the separation of the regulatory from the managerial functions of the state was approved in 1997 but has not yet been implemented. The operational coordination of power generation and transmission systems as well as the technical supervision of the high-voltage grid will be the responsibility of the National Load

Dispatch, which answers not to the national electric power utility (UTE) but to ADME (Wholesale Electricity Market Administration). A Wholesale Electricity Market and the Wholesale Electricity Market Administration are established by law, and enable competition between generators who supply this wholesale market, thus giving way to private sector involvement in power generation. Transmission and distribution are kept as monopolies, but UTE is entitled to grant service-area concessions to private agents. Free access to the transmission grid is also established. With the potential for new hydropower projects practically exhausted, and energy demand ever increasing, the government strategy is focused on two fronts: the import of Argentine natural gas to supplement national supply while keeping costs down, and reforms in international energy trade, first in the electricity sub-sector and later in the oil and gas sub-sectors, to facilitate access to cheaper imported energy sources. The new regulatory framework for the power sector, which resembles that of Argentina, was justified by the need to facilitate access to the Argentine power system. Grid interconnection with Brazil is also under way.

Privatization initiatives in Uruguay suffered a setback in 1992 when a trade-union initiated plebiscite revoked, by a 72 per cent majority, a law that would have allowed the sale of public enterprises to private investors. The plebiscite demonstrated the overwhelming public opposition to the privatization of public companies and restricts future proposals for privatizing the assets of UTE (electricity) and ANCAP (oil). Facilitation of private-sector participation in the energy sector is now being pursued via legal loopholes and the introduction of new regulatory frameworks. Private sector participation will be ushered in through public work concessions for essentially monopolistic projects such as the Buenos Aires–Montevideo gas pipeline; through participation in competitive activities such as power generation; or through associations with public companies for new projects. Meanwhile, UTE and ANCAP retain an important role in the absence of any attempt to improve their management. Plans to deregulate the fuel market open the possibility for competition to be introduced in power generation. Meanwhile, the government seeks to take back its regulatory function from the jurisdiction of public utilities, and regain authority over large public works and the entrepreneurial structure of the sector.

The state-owned enterprises of the Uruguayan energy sector operate as collectors of indirect taxes, a fact that has led to high fuel and power prices and tariffs. UTE and ANCAP provide revenues to the state budget and, in general, finance their own investments. Due to this fiscal burden, fuels in Uruguay are among the most expensive in Mercosur. In the electricity sector, differential tariffs imply cross-subsidies. Uruguay's industrial consumers pay the lowest tariffs nationally and regionally. Medium- to low-scale residential and commercial energy consumers pay medium tariffs compared with the region's average, while residential and commercial consumers with relatively high consumption levels pay the highest tariffs. The least privileged consumers in poor neighbourhoods are subsidized. In the context of growing competition and the influence of private entities, it is anticipated that the investment of electric power utilities will be decreasingly allocated to rural, low-income or high-risk areas. In addition to growing geographical competition for energy investments, competition between the energy sub-sectors may spell negative impacts on redistribution, since enterprises in each will devote most efforts to high-income areas.

Lack of Official Concern for Sustainable Energy

Government discourse on the new energy orientation focuses on the significance of increasing energy consumption as a development indicator, and increase in demand is considered a positive sign. In their public declarations, authorities do not show concern for the efficient use of energy or for the improvement of energy-intensity indicators. There are also no incentives or programmes for improving energy efficiency. This is partly due to inadequate resource allocation. It is certainly due to lack of environmental awareness and understanding, on the part of the enterprises, that saving energy can reduce costs.

In the oil sector, the main environmental impacts result from:

- the location of Uruguay's only refining plant in a residential area of Montevideo, on Montevideo Bay, although environmental considerations have only recently been included in design and operational standards for equipment. Oil refining is largely responsible for contamination of the bay, for gaseous emissions and the risk of accidents in the urban area;
- the risks of oil and derivatives spillage in maritime transportation, evidenced by the accident of the oil-tanker *Santa Maria* in February 1997;
- the illegal washing of tanks by oil-tankers on the Rio de la Plata and around the oil buoy of José Ignacio in the Atlantic Ocean; and
- oil refining, with estimated annual emissions of 230,000 tons of carbon dioxide and 490 tons of nitrous oxides (NOX).

Excluding the impacts of the hydroelectric plants built long ago and despite the efforts made by UTE to reduce its environmental impacts, many local problems persist in the power sector. These include emissions of burning gases in the thermal plants Batlle, also located near the bay of Montevideo, and La Tablada, in the city's north-west. There are plans to convert these plants to natural gas in the near future. Substitution of gas will mean a relative reduction in gaseous and particulate emissions but overall emissions will probably grow in absolute terms since it is expected that both plants, particularly La Tablada, which operate only occasionally now, will soon operate with greater continuity. Public controversy surrounded the construction in 1991 of the thermal power plant La Tablada only 150 metres from a housing complex. Local residents, businesses, environmental NGOs and students opposed it. In addition to the usual health risks posed by power plants in urban areas, people were concerned about the potential impacts of high-voltage power lines in such close proximity to people's homes. In the gas sector, it is primarily the risk of accidental explosions at a plant in downtown Montevideo that concerns residents and trade unions following some minor explosions that have occurred over the last two years. It is envisaged that the introduction of natural gas from Argentina and the effective dismantling of gas installations and tanks may alleviate the potential for explosions. However, the import of natural gas presents other significant environmental problems, of which potential leakage is the most prevalent. Second, in the case of Montevideo, natural gas supply is linked to the construction of an underground reservoir in the Santa Lucia river basin. This river provides clean water to Montevideo and is in the major horticultural production area of the country. The technical feasibility of this reservoir has not yet been established, but parliament is already considering a law to authorize its construction. Third, an expected three million cubic metres/day of natural gas will enter the country and only a small proportion of it will replace other fuels. Hence carbon dioxide (CO_2) and nitrogen oxides (NOX) as well as methane (CH_4) emissions will increase significantly. Over 60 per cent of Uruguay's power is already derived from imported oil, half of which is used in transportation, and a further 50 per cent increase in fuel consumption is predicted in the sector by the year 2005.[3] Emissions growth from oil combustion still dwarfs those from other energy sources. Since Uruguay has no legislation requiring the use of unleaded gasoline or gasoline without catalyzers, air pollution is significant, at least in Montevideo.

The transport sector is responsible for the highest level of gas emissions in the country, with an annual estimate at 1.8 million tons of CO_2, 127,000 tons of CO and 23,000 tons of NOX. Serious water contamination is caused by pouring around 20 million litres of waste lubricating oil per year into the sewerage system. There are, further, no provisions in the emerging regulatory reform process for improving energy supply to the poorest social sectors. Since the emphasis of public utilities is on profit maximization, it is unlikely that they will budget for supply to non-profitable clients. Efforts will be to improve tariff collection in poor urban areas where people have tapped illegally into the grid. For this purpose, UTE has already implemented metering control.

Scope for Improvement?

Research into alternative energy sources has received no significant financing, and there are no activities for promoting the use of 'alternative' energies, partly because of resources, but certainly also because of lack of political will. While existing studies show an interesting alternative power potential, there is no strong evidence that it can be exploited profitably under current market conditions. During the 1980s, two programmes were implemented for the rational use of energy in industry and for energy auditing within some enterprises. But these were limited to the industrial sector, and no programmes have ever been implemented in other sectors except for a metering control programme aimed at avoiding energy theft in low-income areas. Since so little has been done to promote energy efficiency, Uruguay's potential to realize profitable energy savings is high. Non-conventional energy technologies should not be considered in isolation but should be analysed within an interconnected system that includes solar, wind, biomass and hydropower. Since it is unlikely that the absence of sun, wind, wood, and rain will occur all at the same time, one or more of these sources will always be available. Hence the viability of the total, integrated system should be assessed. No country can afford to ignore environmental variables, in particular, climate change. Studies drawing upon the technical, market and policy environment in Uruguay have found that for low daily consumption, wind generation is not economically feasible. For systems larger than 20 kilowatts, wind generation alone is not feasible, but the diesel–wind alternative appears workable compared with connecting to the 15 kiloVolt network if the line length exceeds 15km and the average wind speed is not below 5.6 metres per second, which would limit its application to only a few cases.[4] These analyses did not internalize any of the anticipated impacts on the environment or communities or national dependency upon imports. The results are relative given the limitations imposed by this partial approach and should not be used to dismiss non-conventional energy sources as non-viable. Even so, the studies did identify some potentially interesting sites for wind generation, mainly in the south of the country between Colonia and Punta del Este. Despite resource variability posing a major hindrance for developing wind power, the results are encouraging, since:

- in the historical series analysed, no long periods without wind were detected;

- using high-performance generators in each area, wind equipment could generate over 50 per cent of installed power capacity for at least 3,000 hours a year; and
- the latest studies performed by the School of Engineering with state-of-the-art equipment show that wind-generation costs can compete with those of thermal power generation.

Another joint study of the School of Engineering and UTE determined which points of the national territory are suitable for small-scale hydro-electric exploitation. Since most of the costs originate from civil works, installation of micro-turbines should be linked with other projects that make use of the dams to obtain an acceptable level of profitability. Thus, small-scale hydroelectric generation becomes feasible if implemented within an overall economic development scheme in which the dam is also used for other purposes such as irrigation, drinking water and flood control. The study shows that dams should be located near the national power grid. Only in the most favourable spots will micro-hydro facilities compete with conventional sources.

Uruguay receives a mean radiation intensity of 400 calories per square centimetre per day, with great annual fluctuation. Hence electricity production per square metre of panel is nearly 200 kWh/year, with a power capacity of around 150 watts per 'useful' square metre of panel. Medium- or large-scale solar application is not economically feasible since, for a photovoltaic system to be competitive, power unit cost would have to be significantly reduced. This would be possible if equipment becomes less costly or if there is a significant increase in the conversion yield of solar panels. Photovoltaic systems are suitable only for residential lighting, radio and television and communication in rural areas. For higher consumption uses, the degree of competitiveness is rapidly lost.

Sugarcane, rice and sunflower husks are the main agricultural wastes used for energy purposes. Studies performed in the rice-growing area of eastern Uruguay reveal an availability of 60,000 tons of rice husk for an arable area of 96,800 hectares in this region. Study results yielded an average rate of 1.2 tons of rice husks per megawatt hour (MWh) of electricity/steam co-generation in a thermal power plant. This implies an energy potential of 50,000 MWh for the eastern rice-basin region, which would meet the harvest needs of the region and generate a 15 per cent surplus. Taking into account that rice-husk energy generation technology requires the use of steam cycles with special boilers, investment costs will be high – even though rice husks have no opportunity cost. Nevertheless, a plant of this kind will soon be built through an agreement between the major rice-mills of the eastern region and the US Trade and Development

Agency (TDA), which helps US firms gain access to foreign development projects.

Following the oil crisis in the 1970s, Uruguay started a programme to use fuelwood as an energy source for industry. In the 1980s, the number of enterprises that used fuelwood for production increased tenfold, and fuelwood consumption accounted for one-third of energy end use in the sector by 1994, even surpassing oil and electricity. For residential consumption, fuelwood accounts for almost half of energy end use. Without reducing the standing forested volume, in other words on a seasonal production basis, the existing Uruguayan forests and tree plantations could yield a total primary energy output per year of around two million oil equivalent tons, which is of the order of the total end-use energy consumption of the country. Forest biomass is therefore a very important resource from a quantitative point of view, and, in the long term, it constitutes a significant energy reserve for a country that is relatively poor in primary energy sources.

The potential for renewable energy sources has not been adequately researched, although its potential for development is of considerable interest. Given Uruguay's increasing dependency on foreign energy resources, and the environmental impacts – at local and global levels – of their production and use, it can be concluded that these alternative sources should receive greater consideration and promotion. It is necessary to consider the complementarity of the various sources and their integrated applications, rather than their individual potential, and associate them – as in the case of hydropower and rice husks – to other productive needs. It should also be considered that the expected increase in both rice and wood production would result in an increase of by-products that could be used as energy sources.

Bank Focus on Expansion

In general, the policies of the MDBs in Uruguay has placed greater emphasis on expanding generation and transmission capacity than on promoting efficient consumption and energy savings, even though greater emphasis on energy efficiency and saving would eventually reduce the need for these generating facilities. The MDBs' stated plans for Uruguay are the same as those for the Latin American region as a whole, namely: to increase electricity supply, efficiency in power generation and end use; to reduce the burden of the power sector on public finance; and to identify and adopt alternative options to mitigate the negative environmental impacts of electricity supply and end use. The main guidelines are: transparent regulation; import of energy services; commercialization and corporatization of public utilities; encouragement of private investment; and subjection of loans to government commitments.

The only loan since the Bank's latest energy policies were produced in 1992 for which data have been made publicly available is the Power Transmission and Distribution Project (UY-PA-8177) approved in 1995 for US$125 million and to which UTE is contributing US$103 million. The primary objective of this project, in which $164.5 million is being invested, is the renewal and expansion of transmission and distribution lines and sub-stations, including improving the national grid system for an interconnection with Brazil. Other less important objectives (US$2.5 million) relate to technical assistance for the reduction of non-technical losses and demand-side management, staff training in the new structure of the sector and strengthening UTE's environmental unit. The focus of this loan is on the Bank's policies for regional integration and efficiency of generation and transmission. The contribution to demand-side efficiency and environmental preservation within the loan is almost negligible, however.

Two previous loans were Power Sector Rehabilitation (2622-UR), signed in 1991 to rehabilitate part of hydro-plant Gabriel Terra (130 MW), and the Power Modernization Project (3221-UR), which enabled the construction of the gas oil-burning thermal power plant La Tablada (230 MW). The latter required a total of US$242.8 million, of which the Bank furnished US$65.5 million. This project seems to openly contradict the new guidelines of the Bank. The new thermal power plant La Tablada burns an oil derivative and emits air and sound pollution in an urban area. For economic reasons, this station had hardly operated since 1994 and was widely opposed by several sectors of the population.

Historically, the IDB has played a small role in the sector. The only two projects for which information is publicly available are the Programa de Transmisión y Distribución Eléctrica (Electric Power Transmission and Distribution Programme) (903/OC-UR), approved in 1995 for US$54 million (plus a local share of US$35 million) and the Reforma del Sector Energético (Energy Sector Reform) (ATN/MT 5276-UR) approved in 1996 for US$630,000 with a local share of US$310,000. The main objectives of the first project are to create the conditions for an interconnection of the Uruguayan and Brazilian systems and to improve the distribution networks in several cities in the country's interior. The project aims at regional integration and supply-side efficiency goals. The second project, funded with a grant provided by the Multilateral Investment Fund (a branch of the Inter-American Development Bank), is to study a proposal for oil and natural gas deregulation and for a draft law on hydrocarbons. The main goal of this programme is to create favourable conditions for private investment in the sector.

Notes

1. This chapter was prepared by Gerardo Honty, Centro De Estudios Uruguayo De Tecnologias Apropiadas, Programa de Energias Renovables, Casilla de Correos 5049, Santiago de Chile 1183, 11200 Montevideo, Uruguay.

2. Respectively these are the UTE, or National Administration of Power Plants and Electricity Transmission, in the electricity sector and ANCAP, the National Administration of Fuel, Alcohol and Portland Cement.

3. Tabacco, B. et al. (1996) *Estimación del consumo en el sector transporte*, Montevideo.

4. These conclusions are taken from feasibility studies on alternative energies developed by the School of Engineering at the University of Uruguay. They take into account technical factors, market conditions and government policy. Impacts on environment, communities and foreign dependency were not considered.

Part III

Energy Policy for the Future

18

Which Way Forward?

The energy sector is currently undergoing dynamic, radical change. In low-income countries the sector is being aggressively 'liberalized' from state control, mainly at the behest of multilateral agencies such as the World Bank, with financial and political support from the governments of the industrialized countries. This is part of a more general international political trend towards increased private ownership and less state control. Inextricably attached to the energy sector reform process are the crises of inequity in energy use, a perceived lack of sufficient capital to meet projected energy demand, and local and environmental degradation – including the inherent demands and opportunities presented by the international agreements on climate change. The overall energy sector reform process is being largely led by the World Bank and other multilateral development banks (MDBs) and is redefining the role of the state – and by extension the role of the MDBs themselves – in the provision of finance and support for the energy sector. The MDBs are moving their attention away from providing direct financial assistance for energy projects, and are instead concentrating mainly on providing policy and technical advice to 'assist' low-income countries to alter their legal and institutional frameworks to allow increased participation of private actors in the energy sector. The reform model, in its 'ideal' form, involves the 'unbundling' of previously state-owned monopoly energy utilities into separate generation, transmission and distribution companies, which are then further split into regional companies, and sold to private investors.

The overall trend towards private participation in 'development cooperation' has caused a significant decline in funds for official development assistance (ODA). Total ODA declined during the 1990s from US\$56.3 billion in 1990 to US\$40.8 billion in 1995, for example, while private foreign resource flows in the form of loans, portfolio equity flows, and private direct investment quadrupled from US\$44.4 billion in 1990 to US\$243.8 billion in 1996.[1] Most private sector investment, particularly for large, environmentally significant infrastructure projects, does not take place without some form of governmental and/or multilateral co-financing, risk insurance and guarantees. Publicly supported Export Credit Agencies (ECAs) are central in this context, yet ECAs lack the environmental and social criteria that restrict – in theory but not always in practice – the investments that are made by multilateral development banks. In addition, ECAs generally refuse to reveal the most basic information – a breakdown of transactions by country, sector, and projects financed and guaranteed – although this information is reported as a matter of course in the annual reports of the MDBs.[2]

There is clear evidence that the shift towards private sector finance of the energy sector during the 1990s has not been accompanied by any significant shift towards sustainable energy. As described earlier in this book, most private investments have been captured by only a small number of countries, and the 'top ten' (Brazil, China, Argentina, Philippines, Indonesia, India, Pakistan, Malaysia, Colombia, and Thailand) accounted for more than three-quarters of all private energy investments between 1990 and 1997.[3] In addition, almost three-quarters of private investment in energy in low-income countries between 1990 and 1997 has gone into constructing new power generation plants using fossil fuels, while the remaining 25 per cent has gone into existing energy utilities, or transmission and distribution projects. More than half of all private investment in energy in low-income countries between 1990 and 1997, moreover, went into 286 new power generation projects on greenfield sites, and only a relatively tiny amount of private investment has been made in energy efficiency or renewable energy.[4]

Access to modern energy services by the rural poor is also being almost totally marginalized, meanwhile, and the World Bank itself has been forced to admit that:

> Liberalizing energy markets, however important, may not be the complete answer. Despite the progress made in encouraging private investment in the electricity industry since the beginning of the 1990s, for example, private companies have shown little interest in extending electricity supplies to rural areas. They have instead preferred to concentrate on more lucrative contracts to generate electricity and to supply industrial and urban consumers. There is evidence, in other words, that creating urban-based energy markets by itself will fail to provide rural electricity.[5]

Against this backdrop of a decreasingly transparent, increasingly privatized energy sector, which is showing no signs of any substantial move away from the use of traditional energy technologies based on fossil fuels (oil, coal and gas) there is now general acceptance that the problem of climate change is real, and serious. While the governments of industrialized countries continue to publicly state their commitment to dealing with the climate issue under the Kyoto Protocol (see Box 2.4) they continue to work with the World Bank, with other multilateral development banks, and with export credit agencies, to directly or indirectly finance the development of energy systems in low-income countries based on fossil fuels, rather than on improving the governance of the energy sector, on opening markets for new and renewable energy technologies, and on promoting the development of rural energy.

The Rhetoric of the World Bank's Sustainable Energy Principles

So which way forward? There is no need to reinvent the wheel. Largely as a result of lobbying from NGOs and others that took place around the time of the Earth Summit in 1992, the World Bank already has clear guidelines relating to sustainable energy, and these are already contained within its existing energy policies. In other words, the Bank already has policies in place that promote sustainable energy. The key sustainable energy principles in the Bank's energy policies are:

A requirement for all power lending will be explicit country movement toward the establishment of a legal framework and regulatory processes satisfactory to the Bank. To this end, in conjunction with other economy-wide initiatives, this requires countries to set up transparent regulatory processes that are clearly independent of power suppliers and that avoid government interference in day-to-day power company operations (whether the company is privately or publicly owned). The regulatory framework should establish a sound basis for open discussion of power-sector economic, financial, environmental, and service policies. The Bank must be satisfied that there is meaningful progress towards this objective.[6]

To gain greater country commitment the Bank will better integrate energy efficiency issues into its country policy dialogue so that they can be addressed at an earlier stage. In the Bank's general country policy dialogue with developing countries, greater emphasis will be given to energy pricing and to fundamental institutional and structural factors that affect supply- and demand-side energy efficiency. The Bank will assist borrowing countries to develop integrated strategies that give consideration to both supply- and demand-side measures. These integrated strategies will also consider unconventional renewable options, incorporate environmental considerations, and include a long-term capital mobilization plan. The energy sector is a candidate for greater attention because of its size, its strategic role in the growth process, and its major environmental impacts.[7]

The Bank will be more selective in lending to energy supply enterprises … Governments should clearly demonstrate that they are putting in place

structural incentives that will lead to more efficient energy supply and consumption.[8]

Approaches for addressing demand-side management and end-use energy
intermediation issues will be identified, supported, and given high-level, in-
country visibility.[9]

The Bank will give greater attention to the transfer of more energy-
efficient and pollution-reducing technologies in its sector and project work.
For all sectors, including basic materials-processing industries, the Bank will
actively monitor, review, and disseminate the experience of new efficiency-
enhancing supply-side and end-use products, technologies, and processes,
and cleaner and pollution-abating technologies as they are developed and
reach the marketplace; help finance their application; and encourage the
reduction of barriers to their adoption. Staff working in all sectors will
explicitly review technology options during project appraisals and in sector
work. Cost effectiveness in pollution abatement is a required goal of policy,
and the best mix of energy-efficient and pollution-abatement technologies
will need to be found.[10]

The Reality of the World Bank's Energy Investments

Despite the above principles for sustainable energy, the main focus of
the Bank's strategy for the energy sector continues to be on increasing
private ownership, and in terms of its actual operations – the day-to-day
provision of loans and policy advice – the Bank's staff are tending to focus
mostly on this aspect of the Bank's energy policies. In particular, they are
focusing on the core principle within the Bank's energy policies that:

> The Bank will aggressively pursue the commercialization and corporatization
> of, and private sector participation in, developing-country power sectors.[11]

As described in Chapter 2, the Bank has been promoting privatization
since the beginning of the 1990s, and has argued that this will lead to the
promotion of sustainable energy. But there is no evidence to date that this
is taking place. Between 1980 and 1997 only 1.4 per cent of the Bank's
total lending was for renewable energy, and this rose in the Bank's 1998–
2000 pipeline to only 3.6 per cent. The International Finance Corporation
(IFC – the Bank's private lending arm) became increasingly involved in
lending for the power sector during the 1990s. But only about 7 per cent
of the IFC's energy lending was for renewables during that period; the
IFC has made few investments in energy efficiency-related projects and
has concentrated mainly on investments for the exploration and development of oil and gas, and diesel and gas generation projects.[12] In addition,
private companies are generally not showing an interest in rural energy,

and are instead concentrating on more lucrative contracts to generate electricity to supply industrial and urban customers.

This is not to say that privatization is necessarily better or worse for sustainable energy. The impact of privatization depends on a number of factors, such as whether the country is a net energy importer or exporter, on how many people in the country are connected to an electricity grid, on whether the country is primarily industrial or primarily rural and agricultural, and whether state-owned energy companies have previously been operating at a profit or at a loss. The Bank is not taking these factors into account, however, in deciding whether or not to privatize. Rather, it is promoting privatization as a panacea, which it claims will improve the technical and managerial performance of energy utilities, improve energy efficiency, increase investment, and reduce energy shortages.

The Bank recognizes that public utilities are often not regulated well in low-income countries. This is one of their reasons for privatizing. But regulation is failing generally speaking because legal provisions are too weak to control vested interests in government, economically powerful consumers, and managers in energy utilities. And the Bank is not ensuring that regulation is improved when privatization goes ahead. So privatization is generally going ahead coupled with bad regulation, and this is tending to make the situation worse than before. The result is, increasingly, a privately owned and badly regulated energy sector, which is even less open to public scrutiny than before private sector participation. Meanwhile, in practical terms, it means that a good-quality, but expensive, service is being created for the few in low-income countries who can afford it, while environmental and social goals are falling by the wayside.

As privatization is taking place, larger companies are tending to benefit, local enterprises are suffering, and low-income people who were being served by public utilities are losing much-needed services. During privatization, the accent is being put on cost effectiveness, with the disadvantage that substantial parts of the population in many countries cannot afford power and are losing access. This is not to say that privatization itself is wrong; the problem is rather the indiscriminate and universal application of privatization as an end in itself, which is taking place, rather than the selective application of privatization as a means to promote sustainable energy.

The World Bank is quite right in arguing that many state-owned energy utilities should be reformed because they are poorly managed and suffer from corruption and patronage, preventing a transition to sustainable energy. Simply moving energy utilities from public into private hands, however, does not solve the problem of unjust, corrupt and secretive energy sector management. It is the creation of open, transparent and effective governance that is the key to solving these problems.

The energy sector's size and strategic role in a country's economic development make it a candidate for special attention from the Bank, which should be making effective regulation of the sector a priority whether or not it is privatized. This would be in line with its existing policies, but the Bank is instead aggressively promoting privatization even when the conditions for good regulation are not in place.

Browsing through the Bank's publicity material, or visiting the World Bank's website, one is given the impression that the Bank is a major advocate for sustainable energy. But the reform agenda that is being pursued by the World Bank and the other MDBs is not leading to any substantial increase in the use of new and renewable energy technologies, nor is it leading to an extension of energy services into rural areas.

Fuel for Change

The Bank's latest Energy and Environment Strategy Paper, *Fuel for Thought*, has been put forward as a document that will lead to a shift in the Bank's investments towards sustainable energy. It certainly contains a wealth of ideas. But the number of new sustainable energy projects promised in *Fuel for Thought* remains small in relation to the Bank's total portfolio, and the paper reasserts the Bank's insistence on privatization of the energy sector as an end in itself, with a major focus still on lending for coal, oil and gas.

During the process of producing *Fuel for Thought*, internal disagreements took place at the Bank that led to compromises in the document's contents, making it big on rhetoric but weak on action. The general inadequacy of *Fuel for Thought* as a strategy to promote sustainable energy, as well as the internal disagreements within the Bank that took place during its production, are noted in the United States government statement on the issue, which was issued by the office of the US executive director to the Bank in 1999:

> Unfortunately, there are several troubling aspects of the *Fuel for Thought* process. In particular, it appears from our vantage point that despite nearly two years of consultations, operations staff did not focus effectively early enough on the implications of what implementation of the strategy would entail. When they did, implementation issues and difference in perspective across the Bank forced a virtual return to the drawing board ... something is seriously wrong when a strategy paper requires a major overhaul at a late stage ...
>
> The Energy and Environment Strategy (*Fuel for Thought*) falls short in a fundamental respect. Simply put, we have the impression that *the exercise*

did not really change the Bank's way of doing business in the energy sector ...
Clear and ambitious targets for renewable energy and energy efficiency that
we had called for are missing ... Most importantly, we have the sense that
*the regional action plans state – albeit with precision – activities that were already
in the pipeline, rather than representing additionalities ...*

What can we do at this point except treat *Fuel for Thought* as a living
document and move on to closely monitored implementation? It is clearly
untenable to refer the paper for more revision. [emphases added][13]

Clearly then, *Fuel for Thought* does not match its promise of a change in
the Bank's operations towards sustainable energy. But if *Fuel for Thought*,
the Bank's flagship in this field, cannot change the Bank, then what can?
It is beyond question that the Bank has both the financial and intellectual
resources to produce or obtain high-quality 'thought' on all subjects related
to sustainable energy, be this in the form of policy papers, reports, books,
seminars, discussion papers, or even 'consultations' with 'stakeholders' held
on the Internet or elsewhere. Words are easy, however, while it is the
operationalization of the already copious and mature 'thought' on rural
energy, energy efficiency and renewable energy that is needed. Detailed
principles for the promotion of sustainable energy are already contained
within the Bank's own energy policies. But the majority of the Bank's staff
is largely ignoring these principles and is focusing mainly on a 'one size
fits all' programme of energy sector privatization. Clearly, the Bank's
current attitude to privatization as a 'panacea' must be revisited. The main
focus of the Bank's attention is currently on *ownership* of the energy sector,
while it is *good governance*, in the form of transparency of decision-making,
accountability of decision-makers, and the participation of civil society
and affected people in decisions that affect their lives, that is the key to
promoting sustainable energy.

In terms of specific action in the short term, governments could hold
the Bank's staff accountable for implementing the sustainable energy principles contained within its existing energy policies very simply by ensuring
that these be converted into mandatory Operational Policy, as described in
Chapter 2. But at the time of writing it seems that there is still too much
political resistance from within the Bank's staff, and not enough political
will from the Bank's shareholder governments, for this to take place.

The problem facing sustainable energy is not a lack of ideas. Rather, as
the country studies in this book describe, the problem is a surplus of
vested interests in governments, business corporations, powerful consumer
groups, energy utilities, and, unfortunately, the World Bank itself. The
Bank has already produced more than enough fuel for thought. But if it
wants to be taken seriously as a promoter of sustainable energy in the

twenty-first century then it must prove itself capable of producing real *change*.

Notes

1. World Bank, 'Global Development Finance' tables.

2. Rich, B. (1998) *Export Credit and Investment Insurance Agencies: The International Context*, Washington, DC: Environmental Defense Fund.

3. Private Participation in Infrastructure (PPI) Database, World Bank.

4. Izaguirre, A. K., (1998) 'Private participation in the electricity sector – recent trends', in *Public Policy for the Private Sector*, World Bank, December.

5. World Bank (1996) *Rural Energy and Development: Improving Energy Supplies for Two Billion People*.

6. World Bank (1993) *The World Bank's Role in the Electric Power Sector*, p. 59.

7. World Bank (1993) *Energy Efficiency and Conservation in the Developing World*, pp. 70–1.

8. Ibid., p. 71.

9. Ibid., p. 72.

10. Ibid., p. 74.

11. World Bank (1993) *The World Bank's Role in the Electric Power Sector*.

12. World Bank (1998) *The World Bank Environment Strategy for the Energy Sector: an OED Perspective*.

13. US statement on *World Bank Energy and Environment Strategy Paper* (*Fuel for Thought*), 1999.

Appendix: The Multilateral Development Banks and Energy

The Big Five

The principal Multilateral Development Banks (MDBs) besides the World Bank, are: the Inter-American Development Bank (IDB), which lends to Latin America and the Caribbean and which is based in Washington, DC; the African Development Bank (AfDB), which lends to the African continent and which is based in Abidjan, Ivory Coast; the Asian Development Bank (ADB), which is based in Manila, Philippines, and which lends to the Asian region; and the European Bank for Reconstruction and Development (EBRD), which is based in London and which lends to Central and Eastern Europe and the former Soviet Union. With the World Bank, these institutions make up the 'big five', which are the main official channel for capital resources from industrialized to low-income countries. In addition, there are numerous smaller multilateral lending institutions, such as the International Fund for Agricultural and Rural Development (IFAD), the Islamic Development Bank (IsDB), the Arab–African Development Bank (BADEA), and the Organization of Petroleum Exporting Countries Fund (OPEC Fund), for example, which play an important role but which are much less significant in terms of the total size of their loans. Of all the MDBs the World Bank is the most influential in setting policy for the energy sector, which then tends to be followed by other MDBs. The European Investment Bank (EIB) should also be mentioned since it makes annual loans equivalent to those of the World Bank; the EIB has traditionally focused most of its attention on Western Europe, with some loans also going to areas outside Western Europe such as the Lomé countries, but is increasingly lending to Central and Eastern Europe.

It is beyond the scope of this book to give a detailed account of each of the MDBs. Nevertheless, it is helpful to understand the policies and operations of the World Bank in their broader context. This appendix therefore gives a brief overview of the energy sector policies and operations of the main regional development banks that cover Latin America, Africa, Asia and Central and Eastern Europe.

These MDBs have different backgrounds and face different problems. In the context of this book, however, it is interesting to note the extent to which they are similar to the World Bank. Specifically: they all began to give more attention to private sector participation in the energy sector starting around the early 1990s; they all tend to focus most of their attention on privatization of the energy sector in their operations; they all have policies, or are in the process of producing policies, that relate to sustainable energy; they are each responsible for sustainable energy investments making up only a small proportion of total investments; and they have all been criticized for their lack of attention to sustainable energy and inadequate public participation in discussions on energy policies and investments.

The African Development Bank (AfDB)[1]

In pursuance of its main objectives and in recognition of the importance of energy in the overall economic development in the continent, the AfDB has, since the 1970s, been actively assisting African countries to develop their energy sector. The AfDB recognizes that Africa's energy resource base is rich but not utilized efficiently, partly as a result of institutional and economic structural weaknesses. As is the case with the World Bank, electricity is the main focus of the AfDB's investments in the energy sector. Between 1969 and 1991, for example, the ADB's overall commitments amounted to US$22.3 billion, of which US$2.1 billion, or just under 10 per cent of the total commitments, went to the energy sector.[2]

The AfDB has been assisting its member countries in the selection, preparation and financing of energy infrastructure without a specific lending policy for the energy sector. The bank group has made several attempts since 1990 to come up with an elaborate energy policy but to date, only a draft has been prepared. The draft 'Energy Sector Policy' produced in 1993 has been criticized by many NGOs in Africa for being more rhetorical than concrete and for not adequately addressing the energy and environmental needs of the African countries, and is currently under review by the Bank group.

The draft energy sector policy borrows heavily from the World Bank policy. It states, for example, that the AfDB will support efforts to develop energy demand-side management strategies and will also support member countries in building their capacity to implement an integrated least-cost energy resource planning approach. The policy further states that the bank group will support privatization of publicly owned energy sector companies and adopt the long-run marginal cost concept in its lending operations. The AfDB's energy policy further seeks to encourage both inter-sectoral and regional coordination in all national planning activities and support efforts aimed at improving the environmental performance of supply facilities, especially the encouragement of solar, wind and small scale hydropower. The policy also states that the AfDB will address energy sector activities that take women's needs into account and that it will also support energy sector information systems networks.

Given that over the years the AfDB has been operating without an energy sector policy, it is of course impossible to assess its compliance. The bank has, in recent years, publicly declared its support for new and renewable energy technologies. The bulk of its loans have nevertheless gone to the power sub-sector and only a very small amount of loans have been made for solar energy or biomass projects.

The Asian Development Bank (ADB)[3]

The ADB's Policy for the Energy Sector was approved in May 1995. The policy addresses 'the need for efficient energy supply in Developing Member Countries (DMCs)'. The policy further includes measures to promote energy efficiency and renewable sources of energy.[4] The three main themes of the policy are: defining an appropriate role for government; enhancing efficiency of production, transportation and end use of energy; and more closely integrating environmental considerations in all energy sector activities to enable sustainable development.

The policy stresses that the energy infrastructure of Asian countries has become large, unmanageable and inefficient. The solution to these problems, according to the ADB, is privatization and private sector participation in the energy sector. This involves separating the regulatory, management and ownership functions of the government; restructuring utilities as corporate, commercial entities so they can function with managerial autonomy; enabling utilities to issue equity and bonds and raise long-term borrowings from the capital markets based on their financial performance; and enabling utilities to raise the needed equity to finance expansion through retained earnings. The policies further call for integrated resource planning and the aggressive pursuit of demand-side management (DSM) programmes (see Box 2.3).

The policies make it clear that the exploration and development of fossil fuels (oil, coal and gas) in Asia is 'critical to DMC's economic development'. In this context, the ADB promotes fossil fuels as primary energy sources to help DMCs achieve self-sufficiency or as commodities for cross-border trade.

The ADB sees the private sector as the main engine of structural reforms in the energy sector in Asia, and sees private sector participation plus market-oriented behaviour as crucial to improved performance and efficiency. In relation to this, it divides Asian countries into three broad categories, with corresponding roles for the private sector. These are as follows:

> In DMCs *with very weak institutional bases*, technical functions such as operations and maintenance of power plants, and commercial functions such as billing and collection could be contracted to private firms through competitive bidding.

> In DMCs *with stronger institutional, operational and financial capabilities*, the private sector could be attracted to independent power generation ... turnkey hydrocarbon production and refining and transmission activities, provided a credible legal framework exists.

> In DMCs *with a mature energy sector*, the private sector could be interested in investing in its own or jointly with the public sector, or in making equity investments in energy sector entities that have been successfully restructured into corporations and listed on the stock exchange.[5]

In 1992, at the same time that the World Bank launched its energy sector privatization programme, the ADB created the Private Sector Support Unit within its Private Sector Department, to advise DMCs on privatization and to implement build-operate-transfer (BOT) and build-operate-own (BOO) schemes. The ADB's 1995 *Annual Report* further made it clear that the ADB would continue its privatization drive, based on 'creating a favourable economic and policy environment ... to accelerate private sector development through BOT/BOO projects in the power, roads and telecommunications sectors, and through the privatization of state-owned enterprises'. The 1992 *Annual Report* further stated that 'the main focus of the Bank's work in relation to government-owned energy utilities would thus be on restructuring them into corporations, commercializing them and partially or fully privatizing them. [Moreover] Technical Assistance related sector work activities will be directed to DMCs that are amenable to proceed in this direction.'

The ADB aims to lower energy system losses on the supply side from the present level of around 25 to 35 per cent to an 'acceptable' level of 15 to 18 per cent. This is to be carried out by rehabilitation of existing power generation and transmission facilities, and the introduction of efficient and economic operations and maintenance practices, load management, promotion of technologies and privatization.

On the demand side, the ADB's strategy combines: setting energy prices to reflect long-run marginal cost of supply, border prices and opportunity costs; legislation and enforcement of sound environmental standards; trade regimes and investment regimes that allow the easy flow of energy efficient technology and goods; national and regional efficiency standards for energy appliances and equipment; establishment of testing facilities and the introduction of energy labelling requirements; and tax and other forms of incentives for industries, households and commercial establishments to adopt energy efficient technology and equipment.

Energy pricing is seen by the ADB as a major determinant of energy demand, supply and end-use efficiency and the ADB's goal is to make prices market-driven. In this context the ADB sees the total, albeit gradual, eradication of subsidies in the power sub-sector as a means to encourage competition and to increase energy efficiency.

In relation to the environment, the ADB's 1995 energy policies focus on the need to mitigate the adverse impacts of energy projects, partly through the Environmental Impact Assessment (EIA) system. The policy further states that the ADB, through Technical Assistance activities, will assist signatories to the UN Framework Convention on Climate Change (UNFCCC) meet their commitments (see Box 2.4).

The policy's section on rural energy includes sub-sections on the need to mitigate the environmental impacts of traditional energy systems, and the need to make rural energy systems economically and financially viable. Renewable energy is mentioned in the policy documents only within the section on rural energy and only as options for 'remote locations in isolation', or in combination with other supply sources. Again, the ADB expects the private sector to market these systems, while governments are expected to adapt their trade policies to enable import, manufacture, marketing, and servicing of renewable energy systems.

The ADB's energy policy recognizes integrated resource planning as an invaluable tool in energy planning (see Box 2.3). Nevertheless, its policies promote economic expansion through high-growth industrialization, which of course leads to large increases in energy consumption. In this context, integrated resource planning serves merely to mitigate the negative environmental and social impacts of increased industrial energy consumption.

The European Bank for Reconstruction and Development (EBRD)[6]

The EBRD's Energy Operation Policy of March 1995 contains a number of improvements over the first energy policy, which was introduced in March 1992. The main criticism of the initial policy was its lack of emphasis on demand-side efficiency and its emphasis on supply-side energy efficiency, energy security, and increasing energy exports. The 1995 policies, however, give attention to: increasing energy efficiency and cost effectiveness in both energy supply and demand; supporting and accelerating the establishment of competitive and efficiently regulated energy markets; facilitating integration into international energy markets; improving energy sector environmental performance; and improving the safety of nuclear power plants.

The EBRD's new energy policy is marked by a strong emphasis on energy efficiency. The policies state that: 'the EBRD will invest directly to reduce the energy intensity of demand; develop local production of energy efficiency-related equipment; and develop local financial intermediaries, energy service companies (ESCOs) and third-party financing instruments.'[7] A key element in implementing this strategy has been the establishment of an Energy Efficiency Unit, which, in 1997, planned US$500 million in energy efficiency investments in the CEE region.

The Energy Policy has a special section on nuclear energy, as the EBRD is the only MDB that supports nuclear projects. The Bank operates in the nuclear sector through the administration of the Nuclear Safety Account (NSA) and through so-called Technical Cooperation Funds. Safety aspects for nuclear power plants are also addressed in the EBRD's energy policy: 'Standards applied for construction, management and operation of the plant would have to be fully in line with fundamental principles set out in International Atomic Energy Agency (IAEA) documents' and 'plant safety assessment will be based upon an approach demonstrably equivalent to good Western practices'.[8] This rather vague definition has, however, allowed the application of lower standards than would be applied for the same type of reactors in Western Europe. The EBRD has, moreover, been criticized for its very close links to the nuclear industry and for keeping grant agreements for nuclear projects secret rather than making them available to the public in order to facilitate public scrutiny and debate.

Public consultation is one of the major elements of the EBRD's mandate to promote 'democracy' in the former Soviet Union. A policy on the disclosure of information was approved in April 1996 and became effective in September of that same year. For the first time, information on projects was also made public through project summary documents (PSDs). These PSDs on private sector projects must be released at least 30 days before board consideration, but it is still possible for information to remain confidential at the request of the client or co-financer.[9] Integrated resource plans are, furthermore, not made available to the public for discussion.[10]

The Inter-American Development Bank (IDB)[11]

The institutional structure of the energy sector in Latin America is changing dramatically. The Multilateral Development Banks (MDBs) are among the main agents pressing for these changes, which consist mainly of:

- Privatization of state utilities that deal in energy resource exploitation or generation and/or its transportation, distribution and industrialization. Where privatization is not possible, state utilities are to operate under the logic – and presupposed efficiency – of private enterprise. In some sectors, privatization does not consist in the sale of public companies or assets to the private sector but in concessions to private agents for oil, gas or other kinds of exploration and exploitation.

- Elimination of monopolies and introduction of competition in the energy markets. Competitive logic should replace the conception of public service under which the state guaranteed service with some independence of the purchase capacity of users. Within the new framework, the creation of wholesale markets is fostered, as in the case of electricity where administrative bodies represent generators, transporters and distributors, large consumers and the state. The idea is to create 'open' markets with theoretically free access to new operators.

- De-verticalization or fragmentation of the energy chain into basic functions – of generation, transmission and distribution in the electricity sector – operated by different agents, as a way of promoting specialization and competition and avoiding market 'distortions' such as transference pricing within an economic group or big company.[12]

- New regulations or regulatory frameworks that guarantee free competition in energy markets prevent the formation of monopolies, oligopolies or other 'market distortions' and, supposedly, enable consumers to defend their interests. For this purpose, the creation of regulation and control bodies is proposed.

- New tariff systems – mainly in the electric power sector – to enhance and maintain the financial soundness of companies by absorbing costs and ensuring profits, hence making them attractive to private capital. This implies the setting of market prices and the removal of subsidies to enterprises and groups of users. Under this approach, energy-users are *clients* of profitable enterprises.

- Regional integration through physical interconnections (gas pipelines, oil pipelines, bi-national hydroelectric plants, electric interconnections, etc.) and the establishment of common energy markets to facilitate international energy transactions. The aim is to make the energy systems of different countries complementary and to optimize the contribution of each to achieve cost reduction.

This new model for the energy sector is gradually being implemented in Latin America. It is progressing unevenly in various countries depending on the balance of power between those in favour and those resisting the new model. In some cases, the existence of authoritarian governments facilitated the transition to full implementation of the model. Argentina and Chile are in this sense paradigmatic, or 'examples to be followed' according to the international financial institutions and consultants who profit from dissemination of this model.

The strategy of the Inter-American Development Bank (IDB) pursues a policy similar to that of the World Bank, of encouraging a new institutional and regulatory model for the energy sector. In its paper on energy strategy (*Energy Strategy Profile*, draft, IDB, 26 December 1996), the IDB states:

> The challenge faced by the Bank in designing its strategy for the region's energy sector, is how to optimize the use of its resources to support the countries in achieving their sustainability goals and aims. During the period under consideration, this implies support for the process of transition to a sector ever more autonomous and clean, less dependent on the State and/ or direct intervention by multilateral banks, and with a larger private sector participation.

The document affirms

> the necessity that the Bank's financial support acts primarily as catalyst to attract other financing from the private sector and that its instruments are targeted to ameliorate the risks perceived by investors. Thus, a large portion of the strategy for the sector is concentrated in identifying the new credit instruments and the Bank's role in facilitating the participation of the private sector. These issues, although approached within the wider framework of financing of infrastructure and the strategy for the private sector, still have special connotations for the energy sector.

The IDB considers that the challenges facing all countries in Latin America and the Caribbean are:

1. Consolidation of the regulatory reforms undertaken during the 1990s. According to the IDB, 'the task of consolidating such reforms shall be the main challenge faced by the countries of the region in the energy sector during the next decade … It is not easy to underestimate the complexity of simultaneously performing three innovative transformations almost unprecedented in the world, namely, restructuring of the sector, establishment of a new regulatory framework and privatization. The task is even more complex because, in many countries these transformations should be accomplished within a process of State modernization and structural adjustment, in order to lay the legal and institutional framework required for the proper operation of a market economy; also, the liberalization of the hydrocarbons market is still resisted despite its

evident benefits for consumers; and finally, energy pricing does not reflect the opportunity cost since it is highly dependent on political circumstances.' (*Energy Strategy Profile*, draft, IDB, 26 December 1996)

2. The incorporation of foreign and national capital investment under reasonable conditions for supporting the expansion of the sector.
3. The development of production and consumption patterns compatible with environment preservation.
4. The extension of modern energy options to the whole population.
5. The integration of regional energy markets.
6. The creation of a truly integrated energy market with an active participation of the private sector, which is hindered, according to the IDB, by the 'trend towards self sufficiency and the protection of state monopolies' that still exist in the region.
7. The inclusion of multi-sectoral considerations in dealing with the energy sector.

While the IDB claims environmental and social justifications for its policies, its performance shows a clear bias toward certain objectives – those that are a part of the new economic paradigm: privatization, liberalization, etc. The pressures exerted by the MDBs on governments and the resources they provide are focused mainly on the modification of regulatory frameworks and the creation of conditions favourable to private involvement, not on social, environmental or energy end-use efficiency considerations.

Intervention of the MDBs in tariff and price setting – and associated regulations – has mainly served to encourage private investment. Such pressures are evident in loan negotiations and conditioning and also in the threats of loan suspension, which are occasionally carried out. Mexico and Colombia are cases in point.[13] Their MDBs' investment portfolios show their priorities. Several projects are oriented toward the proposed changes.

• The World Bank granted a US$46 million loan to the Mexican government for Infrastructure Dis-incorporation – a euphemism for 'privatization'. This project, approved in 1995, was designed to assist the Mexican government in the articulation of policies and strategies and in the preparation of laws and regulations for the infrastructure of three sectors: energy and secondary petrochemicals (33 per cent of the funding), telecommunications and transport.
• The IDB approved a loan of US$1.5 million from the Multilateral Investment Fund (MIF) in 1996 to support the Mexican government in the 'establishment of a regulating framework suitable for private involvement in the natural gas sub sector and the institutional strengthening of the Energy Regulatory Commission' (CRE).

In 1996, the IDB approved a loan 'Support for Privatization' for a total of US$33 million (US$12 million from the IDB and the rest from other sources). The loan is to assist the privatization or concession of infrastructure in the transport, energy, telecommunications and sanitary service sectors.[14]

Notes

1. This section has been provided by Ong'wen Oduor, EcoNews Africa, Nairobi, Kenya.

2. Karakezi, S. and G. A. MacKenzie, (eds) (1993) *Energy Options for Africa: Environmentally Sustainable Alternatives*, London: Zed Books.

3. This section has been provided by André Ballesteros, Legal Rights and Natural Resources Centre, Manila, Philippines.

4. Asian Development Bank *Annual Report* 1995.

5. Ibid.

6. This section was provided by Petr Hloblil, CEE Bankwatch, c/o Energy Efficiency Program (PEU), Kratka 26, 100 00 Praha 10, Czech Republic.

7. EBRD (1995) *Energy Operations Policy*, March p. 17.

8. Ibid., p. 26.

9. EBRD (1996) *Policy on Disclosure of Information*, June, p. 5.

10. Personal communication with Mr Murphy, head of the EBRD Environmental Unit, EBRD Annual Meeting, April 1997.

11. This section has been provided by Ricardo Carrere, Third World Institute, Montevideo, Uruguay.

12. Transference pricing refers to prices used in intra-firm transactions between enterprises belonging to the same economic group. Lower or higher prices with respect to market prices permit the transfer of revenues from one company to others of the same group, according to the interests of the owners, and sometimes to the detriment of the minority shareholders.

13. After 1974, some political problems arose between the bank and the Mexican government in reference to the financial management of the CFE. This resulted in a complete withdrawal of the support from the power sector, which lasted almost fifteen years. In Colombia, a serious problem arose between the WB and the government due to discrepancies on tariff levels, which led to the cancellation of the last disbursement of a loan for the sectoral adjustment of the power sector (US$75 million out of US$300 million total). The reason for the discrepancy was that the Colombian government did not allow the acceleration of subsidy dismantling in view of the associated social, political and economic impacts.

14. In Uruguay, for example, the 'Reform of the Energy Sector' project was approved in 1996 by the IDB for US$630,000 with local funding of US$310,000. This loan, provided by the Multilateral Investment Fund (MIF, a division of the InterAmerican Development Bank that concentrates on private sector development), is intended for the study of a proposal for deregulating the fuel and natural gas sectors and a draft law on hydrocarbons. The main objective of this programme is to generate favourable conditions for private investment in the sector.

Select Bibliography

Bosshard, P., C. Heredia, D. Hunter and F. Seymour (1998), *Lending Credibility: New Mandates and Partnerships for the World Bank*, Berne, Switzerland: Equipo Pueblo, CIEL, WWF.

Colley, P. (1997) *Reforming Energy: Sustainable Futures and Global Labour*, London: Pluto.

Datye, K. R., S. Paranjape and K. J. Joy (1997) *Banking on Biomass: A New Strategy for Sustainable Prosperity Based on Renewable Energy and Dispersed Industrialisation*, Ahmedabad, India: Centre for Environmental Education.

Environmental Defense Fund and Natural Resources Defense Council (1994) *Power Failure: A Review of the World Bank's Implementation of its New Energy Policy*, Washington, DC, March.

Foley, G. (1990) *Electricity for Rural People*, London: Panos.

Glen, J. D. (1991) *Private Sector Electricity in Developing Countries: Supply and Demand*, Discussion Paper Number 15, Washington, DC: International Finance Corporation, World Bank.

Hankins, M. (1993) *Solar Rural Electrification in the Developing World: Four Country Case Studies – Dominican Republic, Kenya, Sri Lanka, and Zimbabwe*, Washington, DC: Solar Electric Fund.

McCully, P. (1996) *Silenced Rivers: The Ecology and Politics of Large Dams*, London: Zed Books.

Martin, B. (1993) *In the Public Interest? Privatization and Public Sector Reform*, London: Zed Books.

Mistry, P. (1995) *Multilateral Development Banks: An Assessment of their Financial Structures, Policies, and Practices*, The Netherlands: FONDAD.

Patterson, W. (1999) *Transforming Electricity*, London: Earthscan.

Rich, B. (1994) *Mortgaging the Earth*, Boston, MD: Beacon Press.

Sustainable Energy and Economy Network (Institute for Policy Studies, USA), International Trade Information Service (USA), Halifax Initiative (Canada), Reform the World Bank Campaign (Italy), *The World Bank and the G-7: Charging the Earth's Climate for Business: An Analysis of the World Bank Fossil Fuel Project Lending since the 1992 Earth Summit*, June 1997.

World Bank (1988) *Clear Water, Blue Skies: China's Environment in the New Century*, Washington, DC: World Bank.

— (1993) *Energy Efficiency and Conservation in the Developing World*, Washington, DC: World Bank.

— (1993a) *The World Bank's Role in the Electric Power Sector*, Washington, DC: World Bank.

— (1996) *Rural Energy and Development: Improving Energy Supplies for Two Billion People*, Washington, DC: World Bank.

— (1999) *Fuel for Thought*, Washington, DC: World Bank.

Index